ANCIENT GREEK
GADGETS AND
MACHINES

ANCIENT GREEK GADGETS AND MACHINES

𐌸𐌸𐌸𐌸𐌸𐌸𐌸𐌸𐌸

ROBERT S. BRUMBAUGH

GREENWOOD PRESS, PUBLISHERS
WESTPORT, CONNECTICUT

Library of Congress Cataloging in Publication Data

Brumbaugh, Robert Sherrick, 1918-
  Ancient Greek gadgets and machines.

  Reprint of the ed. published by Crowell, New York.
  Bibliography: p.
  Includes index.
  1. Technology--History--Greece. I. Title.
[T16.B87 1975]   609'.38   75-3983
ISBN 0-8371-7427-9

Copyright © 1966 by Robert S. Brumbaugh

All rights reserved. No part of this book may be reproduced in any form, except by a reviewer, without the permission of the publisher.

Originally published in 1966 by Thomas Y. Crowell Company, New York

Reprinted with the permission of Thomas Y. Crowell Company, Inc.

Reprinted in 1975 by Greenwood Press, a division of Williamhouse-Regency Inc.

Library of Congress Catalog Card Number 75-3983

ISBN 0-8371-7427-9

Printed in the United States of America

CONTENTS

	Preface	ix
1	A Gadget Admirer Abroad	1
2	From the Land of Daedalus	15
3	The Age of Inventor-Heroes	22
4	Scientists and Model Makers	31
5	Mechanical Marvels	46
6	The Democratic Lottery	59
7	The Amazing Archimedes	75
8	Heron of Alexandria	92
9	Automated Toys and Theaters	113
10	The Inventive Instinct	130
	Annotated Bibliography	137
	Index	147

ILLUSTRATIONS

Design for Steam Engine by Heron of Alexandria	4
Coin-Operated Slot Machine	7
Bird That Gives Divine Sign on Top of Shrine	11
Deluxe *Kottabos* Set from Vase Painting	13
Ornamental Fountain with Singing Birds and Owl	17
Automatic Trumpet Sounded by Compressed Air	20
Bird Which Operates like a Siphon	24
Machine to Produce Sound of Thunder	28
Map of the World by Anaximander, Sixth Century B.C.	33
"Dutch-treat" Wine Vessel	36
Map of Agora as It Appeared in the Second Century A.D.	39
Design for Producing Sound on Opening of Temple Doors	41
Self-Opening Temple Doors	44
Revolving Platform	48
Wheel Design for Right-Angle Turn	51
Detail for Right-Angle Motion of Automaton	54
Axle-Winding Patterns for Forward and Reverse Motions	57
Self-Trimming Lamp	62
Kleroterion Lottery Machine of the Fourth Century B.C.	66
Town Clock of 350 B.C.	69
Hercules and Dragon Toy	72

Design for Circular Motion of Automaton	76
Heron's First Automatic Theater	80
The Dancing Bacchantes on Turntable	93
Computer from the First Century B.C.	96
Second Automated Theater	99
Peg-and-Lever Mechanism to Produce Hammering Effect	103
Three Uses for Heron's Steam Boiler	109
Axle Design for Complex Motion of Automata	118
Leaping Dolphins from the Play *Nauplius*	123
Bolt of Lightning Stage Effect	128

PHOTOGRAPHS

Map of Athens, Showing Acropolis and Agora	83
Pottery Grills	84
"Novelty Pitcher" That Held Oil for Rubdowns	84
Wine Jug in Shape of Woman's Head	84
Wine Psykter and Bowl	84
Sphere Supported by Steam	85
Heron's Model of the World in the Center of the Universe	85
Reconstruction of Heron's "World in the Universe"	85
Lead Weight Stamped with Official Seal	86
Standard Measures Used in Ancient Athens	86
Jury Ballots and Tickets	87
Standard Tiles	87
"Tower of the Winds" Clock	88
Death of Archimedes	89
Child's Commode	90
Horse on Wheels Pull-Toy	90
The God Hephaistos on Lamp Relief	90

PREFACE

INDIRECTLY, this book is the outcome of an experiment in interpreting ancient Greek philosophy. A fellowship in Athens in 1962–63 gave me a chance to explore the value of an actual visit to Greece for understanding the history of Western ideas.

During my year in Greece, I kept a set of notebooks in which I jotted down details that I thought might be helpful in getting the sort of picture of ancient Greek philosophy in context that I was after. At the end of the year, when I looked over my notes, I found some ideas that were quite new to me. In the first place, there was an unexpected amount of gadgetry and machinery that I had to make a place for in my reconstructions of ancient Athens and Alexandria: I had previously accepted uncritically the traditional idea that very few of these mechanisms existed in either city. In the second place, there seemed to be a strong possibility of interaction between this technology and the history of classical thought. In particular, some of the items described in my notes could have strongly influenced both the development of science—particularly the atomic theory—and that of Sophistry with its view that values are arbitrary social conventions.

I hope to return to Greece soon, with new notebooks, to continue my experiment in replacing classical philosophy in its concrete context, and to try to prove my hunches about the interaction between technology and philosophy. In the interim, a combination of the interest my friends took in the gadgets I had seen, plus an incurable fondness of my own for gadgetry, have led me to write up this section of my notes for the present book. I have tried to make this account casual and self-contained, and to make it amusing. But it is also a constructive step in a much larger project that I take seriously.

From prehistory in the Near East to contemporary Connecticut, human beings have displayed a restless urge to experiment and invent. Toys, tools, and magic have gone hand in hand with the discoveries of science, with new patterns of political order, new capacities for amusement and leisure.

Even in the brilliant, austere classical world of ancient Greece and Rome, human nature remained the same. A sharp-sighted fondness for toys and ingenious gadgetry existed there, interacting with the development of the "capital of ideas" that Western culture has been using ever since.

For me, rediscovering the magic of the marionettes, the homely hardware of market and kitchen, the old "town clock," the three-wheeled windup chariot makes that ancient Greek and Roman world at once more credible and more real. My interest in "gadgetry" has even led, I think, to some new insights into the world of ideas. At any rate, it has led to some new suggestions that deserve more detailed professional testing. That testing will have to bring together the history of science, the history of ideas, and the inventories of "everyday" sites as more of these are explored by classical archaeology.

Descriptions and diagrams of other gadgets and machines survive in the work of Heron of Alexandria, whose books provide diagrams and descriptions of the Greek scientific tradition up to his time (probably the second century

A.D.). Most of the illustrations included with the text of this book are from two editions of Heron's work. One translation, *The Pneumatics of Heron of Alexandria,* was translated by J. G. Greenwood and edited by Bennet Woodcroft, published in London by Walton and Maherly in 1851. The other is the Greek text, with a German translation by W. Schmidt, published by B. G. Teubner, Leipzig, 1899. Illustrations from other sources are credited where they appear.

Whatever illustrations Heron himself provided (he must have had some, as his text shows) are thought to have been only schematic diagrams of mechanisms. But the text, as it describes animated heroes, gods, dragons, etc., shows that Heron obviously visualized his engines animating artistic figures—probably in the Roman styles of the time.

The illustrations that follow are based more or less exactly on a fourteenth-century Greek Heron manuscript in Venice. Their neoclassicism fits the text beautifully. For the *Pneumatica,* however, a redrafted and relettered version, made in 1851 by a classics scholar hired to do the job for a British professor of engineering, is reproduced. The patron seems *not* to have liked diagrams cluttered with a foreign alphabet, so that all the parts have been labeled with sturdy Roman capitals. For the *Automata,* the figures are from the Teubner edition (Heron, *Opera,* I, ed. W. Schmidt), which are described as close to the fourteenth-century Venice manuscript.

In transmission, figures and text have suffered some damage. Later copyists sometimes did not understand the function of a piece of hardware, or failed to notice crucial mechanical elements as they were copying and embellishing the drawings. But the damage is seldom severe.

English translations of Heron's *Pneumatica* are for the most part from the translation made for Woodcroft; the others, based on the Teubner text, are my own.

Although it may seem, at first sight, out of proportion to

take so many examples of ancient gadgetry from a single enthusiastic Hellenistic author, the character of Heron's work does not really make this so. For he was setting down an encyclopedia or anthology of mechanical devices in a collection spanning the whole period from 400 B.C. to about A.D. 100. A few times Heron himself comments on this long tradition: Philo of Byzantium had almost perfected a fire scene in a play in an automatic theater; Ctesibius had designed a pipe organ with keyboard; and so on. For the most part, though, family resemblances in design and function are the best guide to tracing these centuries of evolution and innovation.

Like Heron's book, this one has some serious comments on the relation of this material both to contemporary images of the Greeks and to interpretations of the history of ideas in the classical period. And, unlike Heron's scholarly prologue, mine has a direct relation to the rest of the discussion. But, like Heron, I have included things that I particularly enjoy just because they interest and please me; and I think any modern reader who shares my affection for this sort of thing will find the illustrations and designs interesting and pleasing, too.

I hope any archaeologist or gadgeteer on tour who sees errors or omissions in my account (Roman padlocks, egg slicers, lopsided Corinthian dice, and similar objects) will be cooperative rather than critical about it, and write to me at the Department of Philosophy, Yale University, New Haven, Connecticut, U.S.A.

My seven-year-old daughter, Joanna, has interrupted my writing of this from time to time to show me her design for a combination quick lunch and Easter hat. The hat is an inverted bowl, ornamented by a spoon in the brim, a box of cornflakes attached on top, and small jars of milk and sugar suspended—in some way she hasn't yet determined —on either side.

PREFACE xiii

This reinforces my belief that from children to adults, ancient Cretan to modern American, human curiosity, with its attendant appreciation of toys and gadgetry, remains much the same. If the reader wants more evidence of this, the books in the bibliography (particularly Britten's *Old Clocks and Watches* and Chapuis and Droz's *Automata*), plus Disneyland, California, supply it abundantly.

In conclusion, the following acknowledgments are made, with thanks, for permission to reproduce maps and photographs and to quote texts. I am grateful to the American School of Classical Studies in Athens not only for my research fellowship there, but for their generosity in allowing me to use so many illustrations from their publications. These are: the drawing of the fourth-century B.C. *klepsydra* in the Agora; the child's commode; the fourth-century horse on wheels; the lamp with picture of Hephaistos; the map of the ancient Agora, as it was in the second century; the pitcher in the shape of a boy athlete; the ceramic grill; the vase in the shape of a woman's head; the fourth-century B.C. *psykter;* lead weights from the fourth century; three standard measures; the official nut measure from the second century B.C.; the jury ballots; and the standard rooftile from the first century B.C. I am also indebted to *Studium Generale* for permission to use three illustrations from the article "Pneuma and the Earth in Space"; to Derek J. deS. Price and *Scientific American* magazine for permission to reproduce the reconstruction of the "Antikythera computer"; to the Harvard University Press, for permission to reproduce Professor Sterling Dow's drawing of the reconstructed *kleroterion;* to the Yale University Library for permission to use the map of ancient Athens in their Cartographic Collection; to the American Geographical Society, for permission to use my reconstruction of Anaximander's map, with coastlines based on Heidel's *The Frame of Ancient Greek Maps;* to the Hafner Publishing Company, for permission to quote from Von Fritz and Kapp, translators, Aristotle's *Constitution of Athens;* and to Thomas Y. Crowell for permission to include quotations from my book *The Philosophers of Greece.* Professor Derek Price, who shares my enthusiasm for gadgetry, has been a friend and helpful colleague for almost ten years. A comparison of his monograph and articles on clockwork and automata with my own text will show how many ideas we have in common, though in general his evidence has come from later periods, and my own finding of their applicability to ancient Greece has been an independent confirmation of them. (An exception here is the classical history of the clock, where I have followed up a series of suggestions he developed in a joint seminar meeting in 1961.)

Finally, I want to thank my family for their interest and help, both on the trip that led to my notebooks on ancient artifacts and during the process of writing them up in relatively clear and coherent form.

CHAPTER 1

A GADGET ADMIRER ABROAD

My great-grandfather Buck was a born gadgeteer and inventor. Two of his patents, an improved spring buggy seat and concrete fencepost, seemed likely to make a family fortune. Unfortunately, he put all his funds into an attempt to construct, for the Patent Office, a working model of his own design for a perpetual-motion machine. Great-grandfather Buck was the only farmer in Polo, Illinois, to whom tinkers were welcome guests rather than itinerant nuisances. My inheritance from this side of the family has turned out not to be a fortune, but a small dash of whatever quirk it is that marks a born gadget admirer. Lacking talent as a mechanic, however, I became interested in the history of philosophy. One patent, issued to me in 1950 for an oriental-language typewriter, is the only qualification I can claim to speak as a professional gadget enthusiast or inventor. But, although I have not invented ingenious mechanisms, I have always noticed, loved, and collected them. On my desk, weighting down a pile of research notes on philosophy, is a chromium combination tool: hammer, hatchet, pipe-wrench, pincers, wire-cutter, screwdriver, and tack extractor.

In 1962–63 I spent a year in Athens studying classical philosophy, and visiting sites and museums. To my surprise I began to notice admirable gadgets and inventions from ancient Greece tucked away without much comment in the dis-

2 A GADGET ADMIRER ABROAD

play cases. My own peculiar fondness for special clay-coil wine jars ("Cool without diluting that delicious Pramnian wine"), improved sausage grills ("New design with lugs keeps your meat from rolling into the charcoal fire"), molds for mass-producing clay souvenir statuettes, and the like, made me unusually aware of the number of these minor displays. Further reading about classical technology supported an uneasy impression I was forming: that the Greeks and Romans, far from being antimechanical and austere purists, were as adept and addicted to ingenious mechanical devices —useful or just attractive through their ingenuity—as many modern cultures are!

When locating the working plans for the first coin-operated slot machine in the West led me to read the *Pneumatica* of Heron of Alexandria (second century A.D.?), I found a set of plans of all the ingenious toys, self-propelled puppets, and pieces of scientific equipment that were surprising enough to use for parlor magic, which six preceding centuries had devised. The reason the old picture of ancient Greece and Rome took no account of gadgetry was not that there were no such items kicking around, but that the ancient Greeks had a common-sense convention (which Heron defied for almost the first time) that these were not the sort of thing people wrote books about. Books were for something else: poetry, philosophy. Gadgeteering was passed on by master–apprentice–patron.

It is true that the ideas of applied science and labor-saving invention were lacking. Neither the notion that mechanisms could be designed for purposes usually thought of as requiring persons, nor the idea of using inanimate sources of power (as opposed to slaves or animals) had made any significant impact. In a prosperous society with slave labor such ideas were not needed. Slaves were an adequate work force for mining, quarrying, road building, sanitation, and so on. The lower middle-class families who could not afford slaves still had no reason for labor-saving devices. With people and animals, extra jobs could be done from time to time with no additional cost or upkeep. Labor-saving inventions would not have been economically useful.

There were inventors, though. Legendary geniuses appeared in folklore and literature, and an occasional historical genius actually applied the ideas of science (Archimedes of Syracuse, in the third century B.C., is the greatest example). Once we give up our instinctive idea that an "invention" must aim at "labor saving," we can begin to appreciate the Greek inventiveness. The ancient Greeks admired machines, with a sort of wide-eyed aesthetic interest, just as a boy admires his twelve-blade knife today. They thus combined skill in designing mechanical things that were useful for their own ends with things they appreciated in their own right.

The mind-set of a modern engineer and inventor always looks toward discovering new power sources and at once tries to find practical application to anything from water purification to space travel. But only rarely does *any* such transfer occur to an ancient "inventor." With a few genuine exceptions, the mechanics of ancient Athens or Alexandria would have loved the idea of great-grandfather Buck's model of a perpetual-motion machine, because the model would have made such a surprising mechanism if it could be built. But it would not have occurred to them that such a machine, built on a different scale from the model, could be harnessed to replace horse or slave power for pumping water, pulling chariots, sawing wood, or waving an enormous fan. Given these differences, a story of ancient gadgets probably comes closer to catching the spirit of some of the inventors, mechanics, and patrons of this early age than more solemn and scholarly histories of technology or of science ever do.

Four themes, in particular, have been used to organize my account. These are self-moving automata and their relation to ancient Greek psychology; astronomical models and their

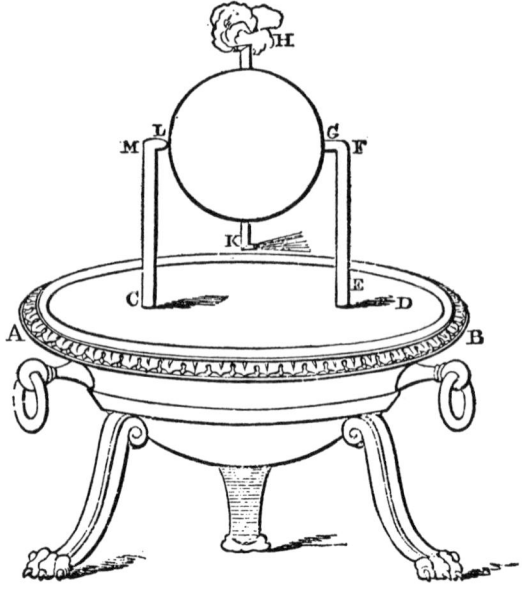

The Greeks invented the steam engine, but to them it was a toy. This design by Heron of Alexandria was accompanied by this description: "A fire is lighted under a cauldron, AB, which contains water and is covered at the mouth by the lid CD. The bent tube EFG extends from the lid to a hollow ball, HK. Opposite G a pivot, LM, rests on the lid CD. The ball contains two bent pipes, connecting with it at opposite ends of a diameter, and bent in opposite directions. As the cauldron gets hot, the steam, which enters the ball through EFG, passes out through the bent tubes toward the lid, causing the ball to revolve."

place in the history of Greek astronomy and cosmology; lottery-machines and secret ballots and their political relevance to Athenian democracy; and the gradual accumulation, primarily in toys for adults, of ideas about the use of power sources that are neither man nor animal power.

Before they invented philosophy and science, the ancient Greeks thought of *psyche* ("soul") as both the vital principal that gave things life, and the inner power of self-motion. "Nature" was alive for them, and each thing that could move did so because it had a "soul" that moved it. The first step toward science, taken by Thales of Miletus in about 585 B.C., still used this older animistic concept to explain change. ("The magnet has a soul, because it moves iron", said Thales.) Could a mechanism move itself, and so imitate the distinctive behavior of living things? And if it could, did this mean that the secret of the creation of a "soul" had been penetrated by the mechanic? From passing references in literature, it seems clear that this question held the same fascination for the ancient world that the synthetic creation of life holds for us today.

The designers of automata seem to have become progressively more ambitious, and their work more admired. Finally, by about the second century B.C. they aspired to nothning less than duplicating the most creative forms of human behavior with their self-propelled series of mechanical components. This idea of duplicating the powers of life by mechanisms must have reinforced the highly speculative thesis of the atomic theory that *all* existing things are complex mechanisms, and led Aristotle, the master psychologist of the ancient world, to agree that mechanical sequences of events in the body were a necessary (though not a sufficient) condition for "intelligent behavior." This project of creating beings with souls mechanically is a strand in the history of Western "gadgetry" which played a part in forming new ideas of the soul, the self, and the difference between nature alive and nature mechanical.

A second goal that men have, and have had for a long time, is to understand the nature of the cosmos, with its alternating seasons and circling stars. In Greece, since about 540 B.C. with the invention of cosmological models by Anaximander, they have felt that construction of a scale model of the cosmos was an important step in learning to understand it, to predict it, even—perhaps—to control it. The simulation of the planets, sun, moon, and stars in their motions was an intrinsically fascinating mechanical problem. Parallel with the designs of self-moving chariots and puppets and with the other ultra-aristocratic current of Greek mathematical astronomy, the tradition of building self-moving universes to scale wound its way into the world of Rome.

An interesting offshoot of this adventure was the display water-clock, in which a sinking float turned an attached revolving celestial globe at one revolution per day. At first as much an aesthetic as a utilitarian object, the sky-globe water-clock put ancient Athens on mechanical clock time by about 360 B.C., and led to use of smaller standard water-clocks which—unlike the earlier sundial—measured time by fractions of a "standard (mean solar) day". The public clock and mechanical time became ubiquitous in the cities of the West, though not in the provinces. In fact an island in the Aegean —Icaria—has not wholly capitulated to it yet. And, in the grand Roman Agora of Athens was built the Tower of the Winds—a beautiful octagonal monument that was a combined weather station and public water-clock. Displaying eight sculptured friezes of the traditional gods of the winds, it was a splendid symbol of the displacement of mythology by science and technology.

Another strand in the history of Greek gadgetry concerns a theme that, so far as I know, has never been studied in detail: the interaction of mechanical inventiveness, measuring tools, and such ethical ideas as democracy, justice, impartiality, and equal opportunity. Justice and the balance, for example, go hand in hand. From the Egyptian court of the dead, through business transactions in the Mycenaean world of the dead, to the official standard weights of the Athenian market, the objective mechanism has been used for operational definition of a most elusive moral quality. I don't know whether there has ever been, or could be, a culture which lacked this interplay between devices and ideals in forming its political and ethical notions; but I doubt if there ever has been so extensive and flamboyant an interaction as there was in ancient Athens in the fourth century B.C. The automation of honesty enlisted impassioned inventiveness in a way that other, more "practical" projects never did—as the Athenians experimented with their newly emerging ideas of impartial trial by randomly selected juries, equal opportunity of every citizen to hold public office, government regulation of local business, and some of the intricacies of foreign exchange.

> Priests were one of the first groups to use machinery for labor-saving purposes. This coin-operated slot machine was used for dispensing sacrificial water in temples. The coin dropped through slot A onto plate at R. The lever at P was then raised and released water at M. The coin slid off as the plate went down, and balance was restored. Water filled the jar.

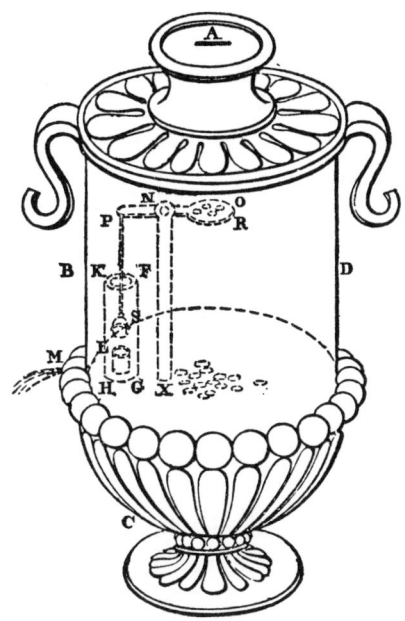

One of the central questions left us by Greek philosophy is: "Can virtue be taught?" We still don't know the answer. Another question we are trying to solve first leaves its traces in the markets and courts of ancient Greece: "Can dishonesty be prevented by foolproof machinery?"

A similar history lies behind American experiments in election procedures, from the paper ballots in a wooden box to the complex tamperproof counters of the Connecticut voting machine.

Business and politics mechanized by standard weights, measures, and election ballots include some things we do not usually think of as gadgetry. But the duel of an idea of objective impartiality with the human impulse to load dice in one's own favor moves from the rudimentary "official measure" into complex and interesting further mechanical adventures that must be considered machines. Even today, there is an element of the gadget in the automatic toll gate that rejects Canadian coins, the pinball machine with a "tilt" circuit to frustrate boys with nickel-steel magnets; and we will meet this same thing in the Athenian Agora.

In ancient Athens, where this same duel between dishonesty and technology had begun, the main targets were short changing by merchants and jury tampering by litigants. In a chapter on Athens, pp. 59–74, I will bring together accounts and archaeological finds that re-create the ancient solutions to the "foolproof mechanization of honesty" problem. We have a pretty full report of the jury-selection devices, since their exact use was specified in the constitution of the city. I am afraid we have a commentary on the experiment preserved in the attitudes toward "useful political inventions": optimism in about 400 B.C., replaced by resigned pessimism by about 340 B.C.

It is sometimes hard to know who is the mother of invention: necessity, laziness, or leisure. For some reason, religion in ancient Crete and Greece and Rome supplied priests who were masters at applying new mechanical principles to their own practical purposes. These practical purposes were of two kinds: sometimes involving performance of ritual with an absolute minimum of priestly effort; sometimes an attempt to impress worshipers and patrons with divine signs and seeming miracles in the temples. In a brief treatment of Crete (pp. 15–21), we will find this tendency at work. It appears again in Alexandria, early in the Christian era, when the first drawings of a coin-operated slot machine appear "such as they use to dispense sacrificial water [in the temples]". In the same source there are plans for "Egyptian machines used to dispense water in temples, upon a wheel being turned." In the Greek religion a ritual washing of hands was a required preliminary to sacrificing by each patron. Without automation the priest would have to be on hand and attend constantly to pouring and emptying the water for this ritual, and collecting the attendant fees. Egyptians, with an anticipation of the modern faucet, solved the first half of the problem: the worshiper could run his own water mechanically, leaving the priest free to concentrate on sacrifice and fee collection. The Greeks, with a brilliant adaptation of an old ballot-box lid design, solved the second part. The coin (five drachmas, representing a reasonable fee) had to be put in the slot to turn on the faucet, leaving the priest free to do as he wished so long as no one stole the whole machine.

The less creditable side of religious automation is also represented in Alexandrian gadgetry, though not in such full detail. The earliest mention I have noticed occurs in the *Mechanica*, a notebook of problems from the third century B.C. In connection with direct-drive wheels, the author says that "this is the mechanism by which the miracles are performed in temples". I wish he had gone into the matter a little further. In the designs of Heron of Alexandria, I have picked up several items that seem to fit the tendency I am tracing. Exhibit A is an adaptation of gearing, at right angles, "to turn birds on a pedestal when the wheel is turned." This looks like an engineered "divine sign" that a gift or sacrifice is acceptable: the worshiper queries the god or goddess by turning the wheel, and if the bird miraculously revolves, the answer is affirmative.

Also, leafing through, I find three more. The first is a use of convection currents to make figures dance "when a fire is kindled on the altar". The second is a use of expanding pressure of hot air to perform an opening of temple doors automatically. The third, a splendid combination of doorbell and burglar alarm, is a trumpet that blares whenever a temple door is opened.

Another interesting feature in the world of ancient machines and gadgets is the interaction of toys and scientific or engineering equipment. This is a subject that historians of science are beginning to treat expertly, and I have some samples that illustrate what they find. One of these is the model made by me and my uncle, Colonel Paul H. Sherrick, who inherited more of great-grandfather Buck's talent than I did. If we are right, magical gadget 45 in Heron's collection was originally a piece of laboratory demonstration apparatus, which proved so unexpected in its operation that it was taken over as an adult toy. And this same laboratory ancestry may lie behind the first use of steam power. Whether it does or not, the steam engine appears as an amusing gadget in Heron's collection, and in a chapter "on the uses of power generated by steam boilers" Heron explains the kinds of toys that steam can operate. To the Greek laymen, scientific apparatus and toys seemed identical. We can tell this from the Athenian comedies of the fifth century B.C. in which part of the amusing effect depends on stage props that caricatured the apparatus of the day.

Our inheritance from the mechanics of ancient Greece has been less influential and less direct than has our inheritance in literature, science, philosophy, architecture, and sculpture.

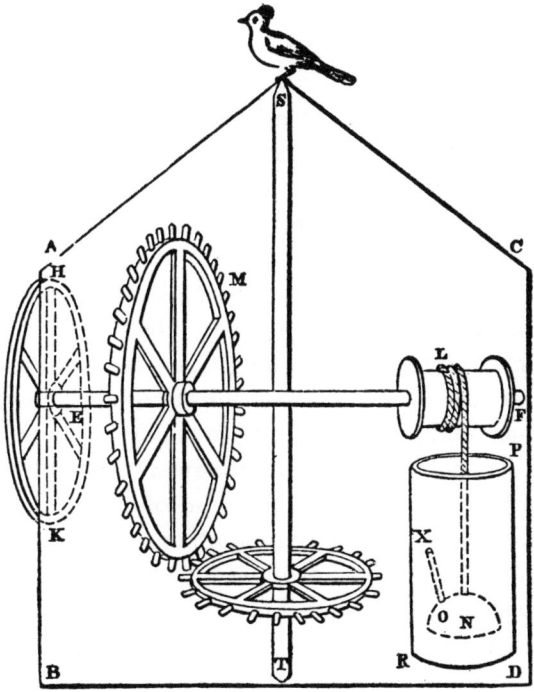

Shrines often contained devices that gave divine signs, indicating acceptance of a prayer or sacrifice. The present bird on its column looks like such a model. If the wheel was turned, the bird would revolve and sing. But the right-angle gearing suggests that the wheels could be disengaged when a favorable response was not wanted. The sound was produced by air pressure.

Greek ideas were sorted out with new notions of what was worth doing by Rome, by the Arabs, by various barbarians. In the absence of literary records, the very existence of machines and mechanics in ancient Greece was forgotten, except for sporadic admirers of Heron's designs. Only with recent work in archaeology and history of science are we discovering the mechanical dimensions of culture in ancient Greece.

The mechanisms were there, and they belong in our reconstruction of the world of the Greek philosophers, artists, businessmen, and politicians. They played a part in the discovery of some great ideas which we have inherited directly: ideas about the self, the universe, honesty, and entertainment.

Did our "many purposes in one" type of gadget also have ancient ancestors? I have noticed two items confirming this point. The first of these appeared in the field of naval weaponry, as the brainstorm of one Stesilaus of Athens. Stesilaus made himself a new weapon—apparently a toothed pruning-hook attached to the end of a long pole—designed to cut rigging of an enemy ship from the deck of his own. It was spear, tree trimmer, and rigging slicer all in one. When Stesilaus first tried this out in a naval encounter, however, his cutter caught in the mast of the enemy ship instead of cutting its rigging. As the two ships sailed past each other, Stesilaus, unable to pull his cutter loose and unwilling to let go of it, raced frantically along the deck of his ship until finally he had to let go and watch his invention sail away. This deck-long dash so amused the other Greeks that the battle was halted: they were helpless with laughter.

The second many-purpose gadget has to be imagined from a tantalizing reference in Aristotle. (Notice, in passing, that traditionally Aristotle and Plato are portrayed as indifferent to technology and hostile to arts and crafts; then notice how many times I will cite them as sources on Athenian ingenuity.) "Nature," Aristotle writes, "is not like the makers of the 'Delphic knife' . . . but each power and organ has its own single function and purpose." In the nineteenth century, when everyone took it for granted that ancient Greece had no gadgetry (and when classicists didn't pay attention to gadgets even when they found them), the passage was baffling. Scholars finally decided that it must refer to cheap knives which had only the edge of the blade made of iron, the rest having been of bronze or some other cheaper metal. Aristotle had meant to illustrate a favorite point of his, that every organ in an animal is built to serve a specific function. If the Delphic knife had really been the cheap model suggested by nineteenth-century scholars, this use of it as an illustration would make no sense at all. Finally, some twentieth-century classicist, perhaps remembering a scout knife from his boyhood, saw the obvious point. The "Delphic knife" has not turned up yet in excavations, but we will recognize it if we find it. Obviously, unlike nature, it was designed for many purposes: use at table, hunting, sacrifice, whittling, probably amateur shoe repair. The Scottish dirk, with two knives and a fork fitting in its single sheath, is another incarnation of the same notion. So is a souvenir my nephew Eric purchased in

The deluxe *kottabos* set on this vase painting depicts a favorite indoor game. Players reclined on couches at equal distances from the target, which was usually a floating copper bowl. Similar to the modern game of darts, *kottabos* was a bit messier—players hurled splashes of wine through the air.

Oskar Seyffert, *Dictionary of Classical Antiquities* (Cleveland: World Publishing Company, 1956)

modern Delphi: a salad spoon and fork, each with a whistle carved into its wooden handle.

Here, again, many-tools-in-one gadgetry has sometimes led to practical applications. The Gemini II flight included in its survival gear a multipurpose rescue light. A compact flashlight, with flashing and direct beams, one side contained fishhooks; on the bottom was a compass, which could be unscrewed; the top had a movable mirror for daytime signaling, with a shallow pocket under it; packaged in an envelope in this pocket was a kit for desalinizing water; in a cylindrical compartment, opened by unscrewing the compass, were stored a long fishing line and a compact sewing kit.

In the following chapters, I have arranged a selection in approximate historical order, running back to about 2000 B.C. The reader who has always taken for granted that Crete, Troy, Athens, and Rome lacked the equipment of modern civilization may be surprised to find that the flush toilet, the safety pin, and the bathtub go back further than the earliest limits of my history. And, in an age when the Soviet Union has recently claimed everyone's inventions as the work of its own culture heroes, it may be amusing to see the ancient Greeks claiming for *their* mythical ancestors the invention not only of self-moving statues but of the alphabet and the wheel!

CHAPTER 2

FROM THE LAND OF DAEDALUS

As THE LEGEND of Daedalus reminds us, before the rise of Greece there was a high civilization on the island of Crete. There the Minoan culture flourished from about 2000 to 1400 B.C. Preliminary to a look at gadgets in Greece and Rome, let us consider briefly this earlier Minoan sea kingdom. It took about two thousand years for the Greeks to catch up with the civilization of these predecessors in the Mediterranean, who often appear as characters in the stories of Greek mythology.

The Minoan world was a gay, civilized one, with an emphasis quite different from the theoretical curiosity of the Greeks or the powerful practicality of the Romans. Plumbing, painting, sculpture, play, religious ritual were the main channels of Minoan inventive expression. Since the palaces were also commercial centers, however, there was some innovation in accounting, storage, and trade. In the course of a visit to the palaces of Phaistos and Knossos, and the Herakleion Archaeological Museum, I noticed a number of items that seemed to anticipate later themes of Greek mechanics and gadgetry. The idea of mechanizing something that one thinks, at first, can be done only by an intelligent performer is reflected in "movable type". The theme of automation of honesty is anticipated—in a small way, to be sure—by standard iron weights and by a tax collector's ceramic model of his town. A mixture of religion, magic, and fondness for novelty is shown in the early "Mother Goddess" pitchers of the Minoan kingdom. An elaborate inlaid gameboard testifies that "toys for adults" had their place in palace life. Libation jars show a genuine desire, on the part of the priests, to use machinery for labor saving (and probably play their part in the history of the clock). An isolated set of quartz lenses is an enigmatic group of "ornamental curiosities", later to develop into the burning glass and magnifying lens. On the other hand, in contrast to later Greek regulation, the standardization is rudimentary and routine. There is no counterpart of the cosmological models of Greek astronomy; there seems to have been no great interest in automata and no development worth mentioning of nonanimate sources of power.

The most impressive modern touch on Crete was the plumbing system in the central palace at the capital city of Knossos. We can still see the system of conduits for running water and for drainage from toilets and baths in the palace. The toilets were like our own, except that the flush tanks had to have water poured into them by hand, not filling automatically. But plumbing comes, I feel, under the general heading of civilization rather than the specific topic of gadgetry.

However, there were gadgets too. I am tempted to think that we *can* count as "gadgetry" the use, in about 1800 B.C., of "movable type". A large disk has been found from that date, with a spiraling inscription in picture writing, a stylized set of pictures earlier than the use of a Minoan alphabet. Perhaps it was a religious or magical disk, one of many pressed from a mold, baked, and sold by travelers in religious supplies. But the striking thing about this inscription is that every picture is *identical* in each of its occurrences. Not just "alike", but absolutely the same, like an original signature on a check and its traced forgery. This can only mean that someone, 3,500 years ago, had invented movable dies and used them to stamp out the writing on the disk or on its mold. Thinking back to my

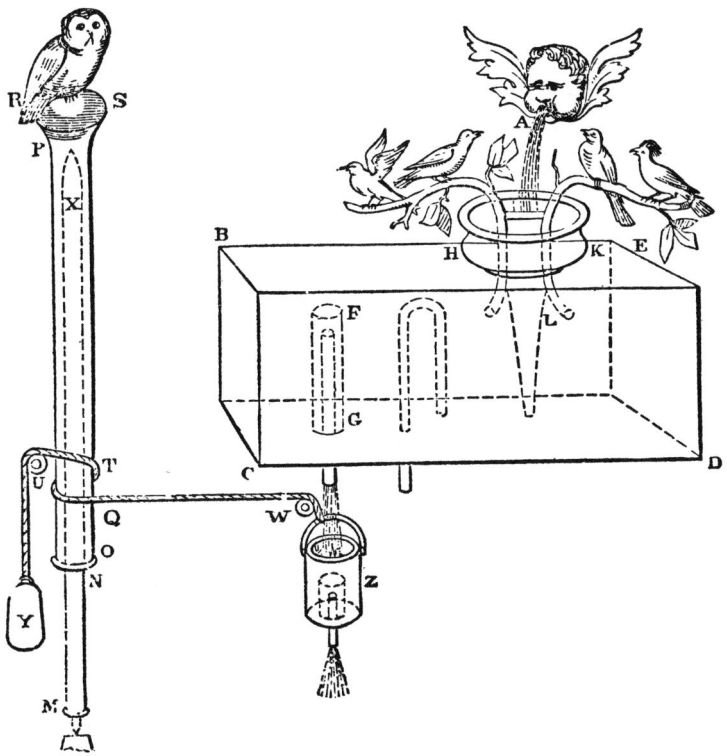

Ornamental fountain with singing birds and owl. The balance between the weight Y and the bucket Z causes the pole on which the owl sits to turn. As the owl turns away from them, the birds sing. As it turns toward them, they are quiet. The bird song is caused by air pressure in the hollow chamber. Yeats, in his "Sailing to Byzantium", refers to trees filled with artificial singing birds of gold, which (according to Chapuis and Droz, *Automata*) were direct descendants of the birds from Heron's fountain.

own adventures in fifth grade, I am morally certain that this inspired idea was the work of an apprentice scribe, remarkable for the messiness of his picture drawing. He must have been very intelligent in other ways to be kept on the job in spite of this handicap (perhaps he could do long division?), and very ingenious to see that once he had his stamp for each picture, freehand drawing of small men, hawks, apples, tripods, and so on would not be required.

A second design, lying somewhere between theology and gadgetry, shows up in the many pitchers shaped like the "Mother Goddess" the Minoans worshiped. The top is the head of the goddess, the pitcher her rotund body. The central feature is an exaggerated pair of breasts, perforated so that through them pour the contents of the pitcher. Such a goddess of bounty, incarnate for the tea table, surely belongs in any catalogue of ancient gadgetry. The "novelty pitcher" passes on to ancient Greece and continues into the modern world. My latest accession in this line is three hollow ceramic Cornishmen, marked "Souvenir of Jamaica Inn, Bodmin Moor".

The important problem of "the automation of honesty" seems to be prefigured, at least, by a find from the ancient Kamares tax bureau. In the collector's office (probably), a large stone slab represented the town: each town building was copied in a miniature flat, baked tile model, properly colored and placed in its correct location on the stone town plan. The red three-story tower, white split-level ranch house, and nondescript gray store still are there in miniature: a reminder of taxes long since paid, or at least long since due.

Probably the most baffling set of items found at Knossos is a case that contains five or six polished, round, slightly convex disks of transparent quartz. They are shaped like magnifying lenses, and do in fact magnify, but their original purpose is not known. Since they would make handsome ornamental stones for attachment to a vest or neckpiece or pendant (like disks of gold and silver that were used for this purpose), it has been thought that their function was purely ornamental. I suspect their attractiveness was partly the way they felt: a smooth, polished quartz pebble is pleasant to touch. But it also seems likely that their magnifying property, unexpected or not, was noticed and that both a new tool and a new gadget derived from the observation. First, the tool: both Greek coin die engraving and some observations in Aristotle's biology seem to presuppose the use of a lens as a tool for magnification. Second, the gadget: in a scene from a comic play, we learn that a "burning lens", made of glass, was a gadget known in Athens in 423 B.C. The "glass" appears as a curious toy—which might be used, it is suggested, to melt the writing from a wax-coated wood tablet recording a legal judgment against one.

A fourth find in Knossos is a gameboard; it counts, I think, as a *toy* in function, but as a *gadget* in design and execution. The board itself is a series of narrow steps—made by inlaying strips of precious stone—leading to three inlaid sets of concentric circles. Three round layer-cake towers, each layer smaller than the one below, are the pieces for the game. Presumably this is a model of fortification and attack which, as the dice fall, moves the towers into position in the circle centers. The board inlay and piece construction, with its craftsmanlike implementation of frivolity, makes me include the game here.

A side case of blocks of metal, each an exact standard talent in weight, prefigures the contest of mechanization against dishonesty of the later Athenian and Roman markets.

My favorite displays from the Knossos palace, though, are the many types and sizes of "libation jar" (*rhyton*) that were found there. Evidently a large staff of priests was maintained to keep the gods appeased by constant offerings of oil or wine upon their several altars. By perforating a small hole in the bottom of a standard storage jar, then setting this directly above the altar after filling the jar with oil or wine, this service could be rendered automatic, and the priests given relative freedom. (Relative because, as far as refilling the jars went, they were already living by mechanical clock time. But in exchange, the hours between fillings were their own, to watch bullfights, solicit funds, or take hot baths.) As I mentioned before, these jars are an exhibit in the history of interaction between machinery and religion, the interaction that culminates in the coin-operated slot machine of ancient Alexandria.

> Automatic trumpet, in the hands of an automaton, is sounded by compressed air. The air is blown down through the trumpet and tube, displacing water in the base. Pressure forces the air back up through the tube, sounding the trumpet.

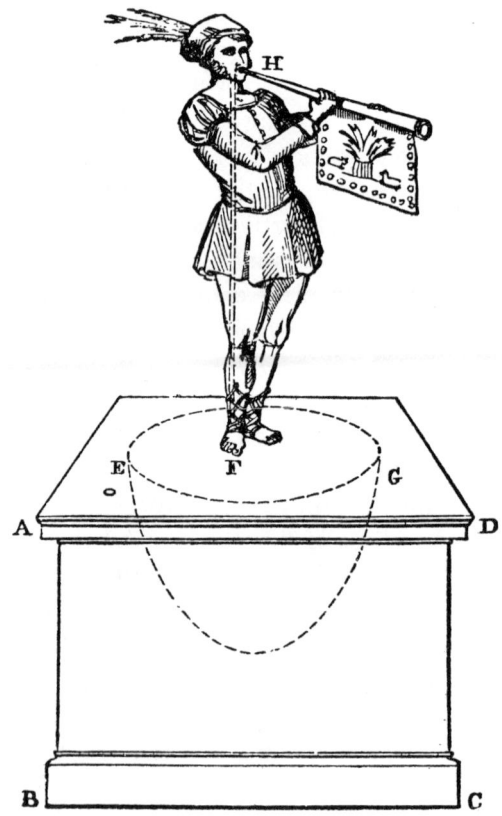

Crete was taken over by Greeks in about 1450 B.C., and the gay Minoan culture was succeeded by the Mycenaean. Sites and finds that I have seen do not indicate any great gaiety in "golden Mycenae". The Lion Gate opens onto a chariot way up to a porch above the Chaos Ravine, where one feels darkness and blood are the backdrop for the shining gold. (The gold cups, death masks, and necklaces were bought in exchange for Mycenaean mercenaries, sold by the battalion to the kings of Egypt.) There were arts and crafts, some of high merit, but no frivolity, and so far as I know almost no gadgetry. One exception may be Nestor's cup, with an ornamental gold dove perched on each handle. For the history of technology, perhaps one should remark that magic made "miniaturization" feasible for the weapons and travel gear buried with the Mycenaean dead. But the age of Mycenae, and of the Dorian invasions that destroyed it, is a dark period in the history of gadgetry.

CHAPTER 3

THE AGE OF INVENTOR-HEROES

In GREEK LEGENDS we hear about three inventor-heroes, who lie somewhere in a shadowy zone between history and mythology. The theory seems to be that in some ancient time the three main cultures of the inhabited world exchanged ideas and techniques. Each hero represents one of the three. Palamedes was given credit for inventing the things the Greeks believed to have been imported from the Egyptians and Phoenicians. Anacharsis was the name of the second inventor, who got credit for discoveries the Greeks believed to have originated with the "Scythians", the barbarians who lived along the Danube River (which the Greeks called the Ister). The third hero, Daedalus, was Greek, a native Athenian who moved—with his own and several stolen inventions—to Crete. This story gives the Greeks credit for the civilization of the Minoans as well as their own.

The earliest-known Greek map of the "inhabited world" shows that these three represent the main divisions. The stories suggest two interesting ideas. The first is the belief that there had been a diffusion of basic tools and techniques between cultures. The second is the idea that every familiar device—from alphabet to ship anchor—must have been the discovery of some single inventive genius. But the stories did

THE AGE OF INVENTOR-HEROES 23

not make their hearers take seriously the notion of deliberately trying to invent useful devices (except for some of the priests, who did). Apparently it was felt that that sort of thing, bordering on magic, could only have happened a long time ago, or only be imagined in a semifictitious story.

We will look in a bit more detail at the three inventors, beginning with Daedalus, the Greek master of all crafts. Daedalus allegedly built "statues that had to be kept chained, so that they did not run away." He became envious of his more talented nephew, who devised the brace and bit and the wood-turning lathe in quick succession, and threw the nephew from the Acropolis of Athens to death on the rocks below. Fleeing from Athens, Daedalus came to Crete, where he designed the palace labyrinth for Minos, the king.

The palace of Minos was three-storied, and on the bottom floor, where more supporting walls were needed, the rooms were smaller and the plan more confusing. To the simple Greek visitors, this ground floor seemed nothing less than a trap; their own "palaces" had nothing like it in complexity. Readers who are familiar with classical myths, or with Mary Renault's novel, *The King Must Die,* will recognize this "labyrinth" as the setting of the tale of Theseus and Ariadne. When Minos and his kingdom were at the height of their power, the king of Athens had to send seven boys and seven girls each year to become bullfighters for the entertainment of the Cretan court. Theseus, son of the king of Athens, went as one of the seven boys. The princess of Crete, Ariadne, fell in love with him and sent him a message to come to her. Theseus had to be guided by a golden thread which was strung through the labyrinth. The two lovers eventually escaped and ran away. Minos was furious and suspected Daedalus of having had a hand in this escape; perhaps he had supplied the idea of the thread. In any case, the inventor and his son, Icarus, were imprisoned on a small, desolate island off the Cretan coast. Here Daedalus showed his

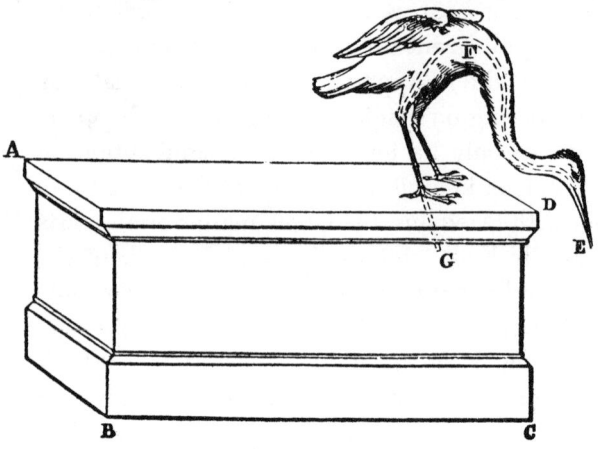

Heron applied the theory of the vacuum to explain the siphon. This is his design for a bird, which will drink any quantity of water presented to it through a siphon extending all the way through its body and beak. The opening at E must be small enough for the water to remain in the tube between demonstrations.

genius; he invented two pairs of wings, made of beeswax and feathers, by means of which he and his son flew away. Daedalus had warned his son not to fly too near to the sun, but Icarus disobeyed his father. His wings melted and he plunged to his death in the sea. When he hit the water, he turned to stone—and I have seen him there, standing alone off the south coast of the island of Icaria, which bears his name.

This legend commemorates Greek respect for their own and for Minoan craftsmanship. The second hero, Palamedes, represents the Phoenician contribution. Son of Nauplius, Palamedes was the inventor credited with both the alphabet and dice. He was put to death during the siege of Troy, accused and betrayed by Odysseus. Another legend traced the alphabet and dice back to an Egyptian god, Theuth. But

Palamedes was given credit by the more patriotic Greek bards, who were, perhaps, reluctant to credit such great inventions to the barbarian Egyptians.

The third inventor, of Greek legend, was Anacharsis the Scythian, from the plains and forests of "Scythia" somewhere to the north and west of Greece. Anacharsis "had designed the first ship anchor, and the potter's wheel."

No one knows what amount of truth lies behind legend here. It *is* true that in the eighth century B.C., the Greeks borrowed a new alphabet from the Phoenicians, having forgotten their own older one. And certainly Daedalus had done *something* remarkable. Perhaps his statues, instead of having feet firmly planted side by side, had one foot advanced for the first time in Greek archaic sculpture. And possibly, as a later poet suggested, the "wings" of Icarus were sails for a small boat. But somehow these explanations seem to undersell the ingenuity of Daedalus. Perhaps someday we will be able to find out more of the detail that lies behind this history of the great Greek inventor. As for Anacharsis, I am at a loss. Neither anchor nor wheel came to Greece by way of Scythia. The Scythians were "red-haired and potbellied, and spent their time drinking and fighting." Their only recorded venture into gadgetry was making drinking cups out of gilded skulls of enemies slain in battle, a practice the Greeks thought in poor taste. So I think more fantasy than history lies behind this third inventor of the folklore trinity.

The wonderful statues attributed to Daedalus come from, and refer back to, a period halfway between the world of science and mechanics and its precursor, a world of magic. The truth about these statues, which "had to be kept chained, or they would run away," has been a real challenge to scholarly ingenuity. In addition to their propensity to run away, there was something remarkable about the eyes of the statues: either they moved or in some other way gave the impression that they could actually see. Unlike the general inventions the

Minoans and Myceneans attributed to Daedalus, the story of the statues suggests some specific and remarkable innovation.

There are three possibilities to be considered: First, the new work of Daedalus may have been magical, not technological. According to this theory the whole thing was pure folklore, based on memories of past magic. Second, it may have been an artistic innovation. Something new in the posture or design of statues made them seem lifelike, and so started the story. The third possibility was that they actually represented something new of a mechanical kind.

In favor of the first of these explanations, it must be remembered that the Egyptians had from a remote past decorated their tombs with "magical" pictures, supposed to come to life as guides and servants for the dead ruler. In Mycenaean tombs, similar magical belief is reflected in the burial of miniature utensils with the dead, including a thin gold-foil balance (to avoid being cheated in transactions in the other world). A great British classicist, John Burnet, thought that the whole Daedalus story was simply an elaboration of this magical tradition. The difficulty with this notion is that the magical tools and pictures were of long standing and widespread use. It is hard to see why they should give rise to such a *specific* story of invention.

The second theory depends on a new technical style in sculpture: representing a figure with one foot advanced, rather than with both feet together, and perhaps painting or inlaying eyes in statues. I suspect that the facts of the history of art go against this notion, at least as far as the advanced foot is concerned.

The third theory raises the question whether or not the statues had some "animating" mechanical innovation as well. Whether they did or not, the legend held out a tempting suggestion for mechanics. By the fourth century B.C., automata had reached a high level of development. Certainly as early

THE AGE OF INVENTOR-HEROES 27

as the second century B.C., a mechanic fascinated by "automation" set out to duplicate the most complex phenomena he could imagine, with a projected mechanical theater.

In the years after Daedalus, Greek philosophy explored the possibility that the *psyche*, the animating force of living things, is itself a material, mechanical part of an organism. This was the thesis of Leucippus and Democritus, inventors of the classical atomic theory. Socrates and Plato reacted vigorously, arguing that human consciousness and behavior require a *psyche* different from, and not reducible to, the body it inhabits and directs. Aristotle then proposed a judicious compromise, in which the *psyche* became the "manner of functioning of a properly organized body" and thus neither separable from nor reducible to the body it animates.

Another inventor-hero of later legend was a Scandinavian who brought a remarkable "magic dart" to Greece in the sixth century B.C. That was the time of a visit by Abaris, a priest from the far north. It was said that in his home country, night lasted for half the year—though most Greeks didn't believe it. Abaris had come to visit the great Greek mathematician Pythagoras, and had brought him a gadget as a present. "Abaris gave to Pythagoras the magic dart which he carried, and by means of which he was able to find his path through the forests." This is the original story; later it is changed to "the magic dart, on which Abaris flew through the air." The magic dart sounds as though some Scandinavian wanderer had been equipped with a magnetic compass over a thousand years before the usually recognized date of the introduction of compasses into the West, and well before the usual date accepted for the invention of the compass in China! We don't know what happened to the amusing magic dart. No practical use was made of it, and no new story of a hero was made up to honor its inventor.

Perhaps we can get some feeling for the plus and minus factors of ancient invention by contrasting the inspired sim-

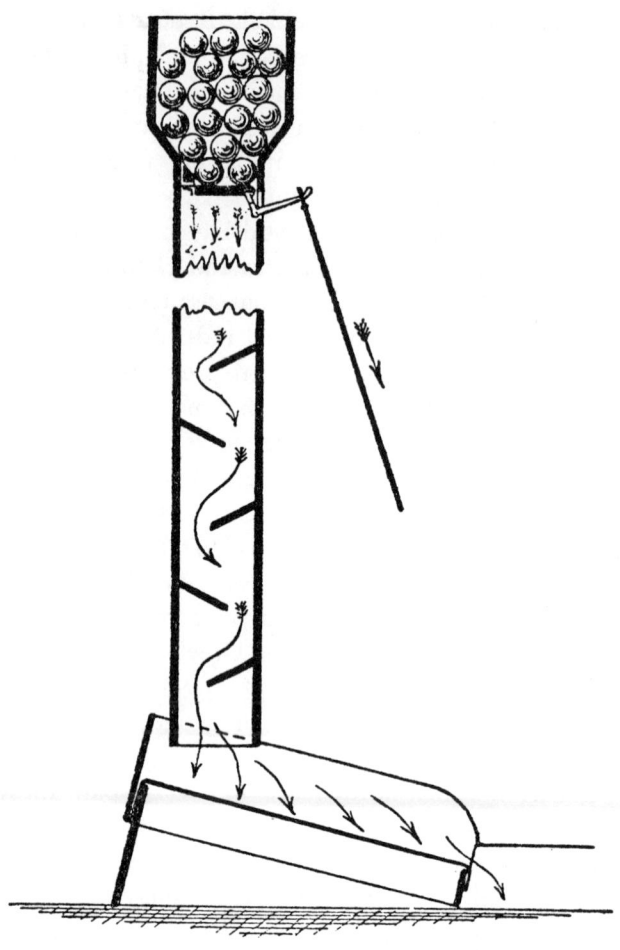

As a sound effect for his automatic theater, Heron invented a machine that produced the sound of thunder. The cord on the right, which was pulled by a motor, released the bronze balls at the top. The balls fell through the baffle onto a tin plate at the bottom. Heron may have been inspired in this design by the ancient Athenian lottery machine.

plicity of the pencil equipped with an eraser to the mechanical imbecility reflected by the early stages of the lock and key. The blunt-ended pencil, or stylus, dates back to the earliest history of writing itself. On clay or wax, "erasure" could be made by rubbing a smooth stylus-end over the tablet surface. The earliest key, on the other hand, was simply a large hook thrust through a hole in the gate or door to catch a latchstring when it wasn't already out. Presently this was improved, so that the hook moved bolts to the side when it was turned. An early philosopher, Parmenides, describes the gate of the House of Day in the underworld in his poem:

There were the great gates, iron bound,
High as aither; and to them, vengeful Justice
holds the double-turning keys . . . .

The fact that this design had any value at all, in the face of the ease with which lockpicks could have been improvised, clearly indicates that in this period, for all the development of arts and crafts, people simply had not learned to think mechanically. Where, as with the stylus eraser, a single glance could reveal design and function, people were appreciative enough; but where, as with the lock, the mechanical logic of a more complex set of moving parts was involved, imagination seems to have followed only dimly, so that childlike security measures still baffled thieves or invaders.

By the Athenian fifth century, the "crackpot inventor" appeared as a recognized character in comedy. We have a scene in which a town planner stakes out lots in the upper air, and one in which apparatus and ideas are for sale in a "Thought Shop". The history of Stesilaus and his new rigging-cutter is another tale in this tradition.

By no means all classical inventiveness centered in Athens. In Tarentum, a city of southern Italy, a fifth-century mayor and mathematician, Archytas, invented a new design of baby rattles. ("Because children should learn music young," according to one report. "To keep youngsters from breaking up the furniture, since young things cannot be still," according to another more prosaic account.) Archytas also invented "a toy bird which actually flew". It seems likely that this was a bird-shaped kite of new design.

At Olympia and Delphi, endless recriminations among chariot racers and fans and judges were ended by the invention of an ingenious cord-operated set of starting gates. Grooves for the string that raised all at the same time have been found in the stone starting lines in the stadia, and the gate reconstructed by Professor O. Broneer.

But it is in Athens that we can best see a mechanical sense and a feeling for gadgetry evolve, which flourishes finally, in practically modern form, in the second-century-A.D. world of Alexandria.

CHAPTER 4

SCIENTISTS AND MODEL MAKERS

WE ARE FIRMLY in history with Thales of Miletus, credited with predicting an eclipse of the sun that took place on May 23, 585 B.C. Thales, by profession an engineer, is the father of both Western science and philosophy. But he was more famous in ancient times as an inventor. "A veritable Thales," people would say, when they were admiring some new invention. Thales' successor, Anaximander, introduced the use of models into science. A star map, a map of the world, and a marvelous model of concentric stovepipe rings of fire to explain planetary motion were among them.

This work by Anaximander of Miletus, in the sixth century B.C., seems to be the first attempt to understand the universe with the aid of a mechanical model. He was the second of three successive Milesian inventors and engineers who, among their other accomplishments, discovered philosophy. Because of the close interaction of science, philosophy, and cosmological models, I would like to introduce the theme here by an account of the Milesians and the Pythagoreans taken from my book, *The Philosophers of Greece*, describing these early schools of thought.

The Greeks thought of Thales as a great inventor, because of his achievements as an engineer. How much they underrated him can be seen from the fact that Thales could, with some right, have claimed the ideas of matter, of physics, of science, and of philosophy as his inventions. However strange this may seem, all of these ideas had to be discovered. And to be discovered, mythology had to be abandoned. To state—as Thales did—that "All things are water" may seem an unpromising beginning for science and philosophy as we know them today; but, against the background of mythology from which it rose, it was revolutionary. The break was not complete; it could not have been. Thales still had no abstract idea of matter, as opposed to an imaginative picture of a fluid sea; the two were mixed together. And his idea of change was still based on a feeling that "All things are full of soul." But he had asked a new kind of question. His question has given distinctive shape to Western thought.

To the Greeks, Thales' concept of a systematic development of natural science made him the great pioneer of thought. But modern scholars are likely to choose for their hero his successor, the more poetic and flamboyant Anaximander. He can truly be called the first real philosopher.

Leaping beyond the brilliant yet simple notion that all things are made of the same stuff, Anaximander showed how deep an objective analysis of the real world must penetrate. . . .

One of Anaximander's great contributions was the general concept of models, which he immediately applied in all the ways we do now. His construction of the first map of the known world shows the same combination of mechanical ingenuity and scientific insight. Very few people realize the importance of models, though we all use them and could not do without them. Anaximander tried to build objects that would reproduce the same operations or relations as the things he was studying but on a different scale of size. One result was a pair of maps: one of the earth, the other of the stars. The map reduces to small scale the distances and directions of places; if we had to

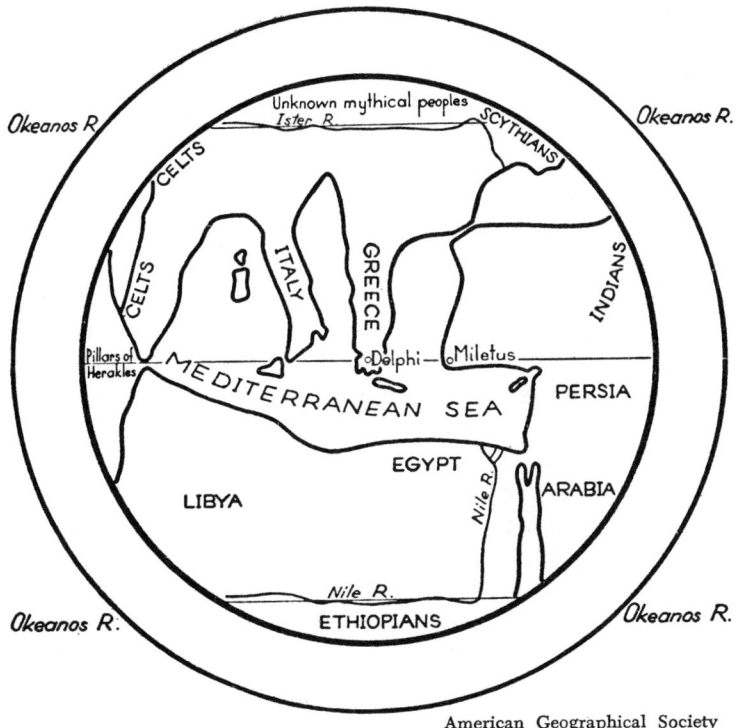

American Geographical Society

This map of the world, drawn by Anaximander in the sixth century B.C., is believed to be the first geographical map ever made. Its neatness and symmetry may be an attempt to make geography reflect the regularity of nature, which Anaximander admired in his astronomy. In the center is Delphi, where a stone—the *omphalos,* or "navel"—marked what was thought to be the exact center of the earth.

depend on travelers' diaries and our own impressions for information about where other states and cities are, travel, commerce, and geography would be in poor shape. Anaximander also built a model of the motions of stars and planets, with circling wheels moving at different speeds. Like our modern planetarium projections, this made it possible to speed up the observed patterns of planetary motion and find in them a regularity and definite ratios of speed. How indebted we are to the use of models can be briefly indicated by pointing out that the Bohr atomic model played a crucial role in physics, and that even a test-tube experiment in chemistry or a rat experiment in psychology is an application of model technique.

The first astronomical model was rather simple and homely; but, primitive as it was, this is the ancestor of the modern planetarium, the mechanical clock, and a host of other related inventions. Anaximander supposed the earth to be a disk in the center of his world, surrounded by rings of hollow pipe (a modern stovepipe gives the right idea) of different sizes and different speeds of rotation. Each pipe is full of fire; but the pipe is made of a hard shell or bark (*phloion*) which keeps the fire inside, except at certain openings (breathing-holes, from which the fire escapes as though blown by a blacksmith's bellows); these openings are what we see as the sun, moon, and planets move across the sky as the circles turn. There are also dark clouds between the wheels and the earth; these are the cause of eclipses, which occur when they hide the openings of the pipes from our view. The whole system has a daily revolution, but in addition each wheel has a proper motion of its own.

Whether the model also treated the fixed stars in this way is not entirely clear. Anaximander seems to have designed a celestial sphere mapping out the heavens, but we don't know how this extension of map-and-model technique was related to the moving mechanism of wheels and fire.

Shortly after the Milesian engineers built their mechanical models of the cosmos, Pythagoras and his followers hit upon a different idea and technique. Convinced that pure mathematics rather than mechanics held the key to science, their "models" were mathematical rather than mechanical. The Pythagorean discovery of "the music of the spheres"—the similarity of simple ratios between planetary motion and concordant lengths of tuned instrument strings—suggested geometrical designs of metal bands arranged in a pattern showing the system's mathematical harmony and simplicity.

A brief quotation from my *Philosophers of Greece* sums up the place of Pythagoras and his school in the history of early Greek philosophy.

> Answering Thales' original question, Pythagoras and his followers held that all things are numbers. His study of the mathematical ratios of musical scales and planets led Pythagoras to believe that quantitative laws of nature could be found in all subject matters. He further expected such laws to have the simplicity of those governing music.
>
> Western thought owes to the Pythagoreans: first, the discovery of pure mathematics; second, a sharpening of the notion of mathematical proof; third, an awareness that form and structure give things their individual identities. Their work initiated the scientific search for quantitative laws and the philosophical tradition of formalism, which finally culminated in Plato.

Perhaps the one device of antiquity which best matches my definition of a gadget is a work attributed to a Pythagorean, Glaucus of Samos. In the sixth century B.C. Hippodamas had applied the Pythagorean discovery of the mathematical basis of music by the construction of "tuned metal disks," the radii of which were in concordant ratio. Impressed by this, Glaucus, famous for his metal working, went it one better. He built a combined trophy, xylophone, and geometrical rod-and-band representation of the musical scale. "A tripod was made by him, which when one struck the knobs and bands, and the rods inside, gave off the notes of a scale. . . ." The tripod, as standard gift or offering, carries religious and aesthetic overtones. This association of the aesthetic and ornamental with the geometrical model of "rods and bands" reflects an attitude that is to prove important in the development of the ancient "public clock". The metal model—combined orrery and xylophone—expresses the discovery that "music is mathematics overheard." The metal-band model of the cosmos has an important later history in Greek astronomy. But the tuneful tripod, which apparently did not find later imitators, remained a one-time curiosity.

In astronomy as in psychology, the proponents of the atomic theory gave the phenomena a purely mechanical explanation. Our world, they held, was only one of many spinning systems of atoms. By its rotation, the atoms in our world were sorted out, with the heavier drifting to the center (as sticks do in a whirlpool), the lighter being pushed toward the perimeter. This vortex model satisfied them that they had

"Dutch-treat" wine vessel empties from one of three chambers depending on how heavy a weight is attached to its outlet. Heron suggests that this may be a good way to see that each person drinks no more wine than he has contributed to the party.

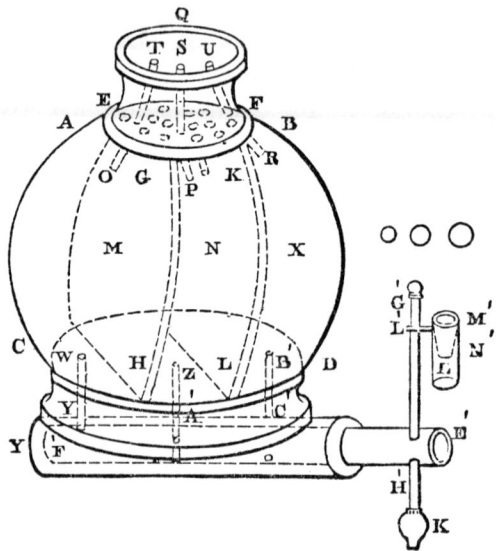

found a purely mechanical explanation for astronomy; neither the Pythagoreans, with their aesthetic-mathematical considerations, nor the widely held opinion that the stars and planets were beings with souls impressed them. As with psychology, so with astronomy, an opposition developed between the philosophical approaches of "formalism" and "mechanism."

In a defense of formalism and idealism against the materialistic atomic theory, Plato and his Academy played a leading role. Because he believed that education and philosophy should be "shared inquiry," Plato wrote in dialogue form, combining elements of myth and drama with abstract argument.

Offhand, one might think of Plato as the last person likely to pay any attention to models or even, perhaps, to experimentation. But the fact remains that in crucial passages Plato does write as though he were visualizing mechanical models: of metal bands, for the planetary orbits; of scales, for calculating relative planetary positions; of some celestial globe, turning "on a fine pivot", capable of reversing its direction; of a colorful mythical view of the universe in cross section.

The standing arguments against the view that the Academy had anything to do with such models have been three. The first is that there are no literary records attesting the kind of mechanical skill that building such models would require. This argument overlooks the fact that literature about technology in ancient Greece lagged far behind actual accomplishment. The second ground for doubt is the thesis that Plato's ideal of pure mathematical science *must* have made him indifferent to or distrustful of observation and simulation. But, while this may seem evident to a modern philosopher of science, I see no reason to think that Plato himself, in the more "catch-as-catch-can" state of research in his own time, saw such a sharp distinction. The third ground for questioning any fondness for mechanism in this school is a denial that the words in the text which seem "mechanical" have any such intended referent or meaning. But philologists are agreeing more and more that such words actually do refer to technological and mechanical processes and things.

If one looks at Plato's writings in the context of what we have seen so far, it seems clear that model design, and probably construction, was a part of the work in the Academy. The most spectacular passage, about which volumes of commentary have been written, is an account in the *Timaeus*, 32 ff., of how God created "the soul of the world". For Plato, the world had a *psyche* which caused its regular motions. What the Creator does, in this account, is to "mix together" the raw materials of existence ("Being, Same, and Other"), "spread them out" in a wide band, then "cut in" a complex set of metric scales. After that, God "splits" the band lengthwise, then "bends" and "fastens" the resulting two bands into rings, one inside the other. Next He tilts the inner ring at an angle, and the soul is set in motion: the outer ring turns rapidly toward the right, the inner, which has been further subdivided, moves more slowly toward the left.

When we read that by this construction the "divine craftsman" (*ho demiourgos*) had completed the "soul of the world", it seems very clear that Plato's account of this creation is exactly a description of what a coppersmith in Athens would have done to build a metal-band astronomical model.

Models play a role in Greek astronomy partly because the early Greek scientists were concentrating on geometrical models of a qualitative kind, which could perhaps *in principle* be adjusted to various quantitative observations. With our heliocentric models we can easily see how and why the planets in orbits outside our own appear to stop and move backward as our earth moves past them; with the geocentric view this observed "retrogradation" was much more baffling to predict or explain. Plato's model of the world-soul, or any model of this type could generate regular spiral paths; but

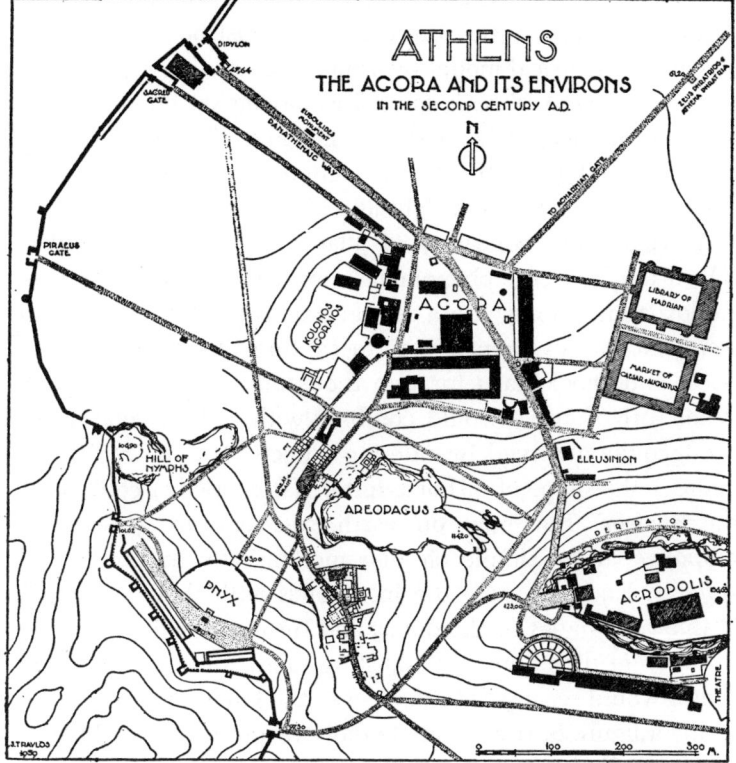

American School of Classical Studies at Athens

The Agora, business center of ancient Athens, as it appeared in the second century A.D. It housed the official "Bureau of Standards", most of the civil courts, and at the west end a temple to Hephaistos, god of arts and crafts. Archaeological finds by the American School in their excavation there have offered many new insights into ancient Greek everyday life and technology.

the planets, still true to their name of "wanderers," kept straying from them. Not only was there periodic retrograde motion, but the planets seemed to "wobble" up and down, so that the observed latitudes did not match the exact predicted spiral paths. Plato mentions these unexplained anomalies in the *Timaeus,* and admits that his model will not explain them. However, he was sure that there must be some simple, elegant geometrical model by which they could be predicted and explained. And he is quoted as having set for the Academy the problem of "saving the phenomena"; that is, finding equations and geometrical constructions that have both mathematical simplicity and a tight fit with observation.

A friend and colleague of Plato's, Eudoxus of Cnidos, solved the "save the phenomena" problem. The solution that Eudoxus hit upon—with a few additions by Aristotle, which were not improvements—was to capture the imagination of poets and theologians for two thousand years in the West. What sort of curve would combine with a daily rotation to produce the observed "wandering" effects? It appeared that a figure 8, the shape of a "horse-fetter" (*hippopede*) could give the answer. But how on earth was one to generate a *hippopede* from simple circular motions? By having *nested spheres,* Eudoxus saw—the inner ones with their axes fixed at different angles in the outer, and turning at different rates in different directions. By adjusting these, an observer in the center would see a point attached to the innermost sphere describe a figure 8. If a complete daily revolution was added to this nest of spheres, as the outermost heavens carry the whole inner system around, the individual figure 8 patterns would be transformed, for our observer in the center, into orbits showing both the retrogradation and the latitude anomaly reflected in the observations.

Whether or not Eudoxus built physical models of his sphere nests of figure 8 curves is not known. But some things about his "saving the phenomena" program are clear. First, his solution must have been intended to be purely mathematical, the solution ending when he could prove that a set of component motions generated the desired curve, in terms of pure solid geometry. Second, however, his solution, for all of its "purity," rests solidly on the long tradition of studying astronomy with the aid of geometrical and mechanical model building.

Complicated and misunderstood by Aristotle, in temperament more a biologist than a pure mathematician, these spheres of Eudoxus, as we have said, were to capture the imagination of the West for tens of centuries. Ironically, they were so satisfying that they held back the further development of purely mathematical astronomy.

Aristotle set himself the task of synthesizing the two ideas of mechanism and mathematical formalism. This led to the invention of a new science of "celestial mechanics" and to the plan of the most complex piece of machinery ever designed (or at least imagined) in antiquity, Aristotle's final cosmological model. In astronomy, as in psychology and the other branches of "philosophy", Aristotle wanted to bring together the insights of the "mathematicians" of Plato's Academy and

Design for producing sound on opening of temple doors. This device could also be used as a doorbell, burglar alarm, or to perform a miracle.

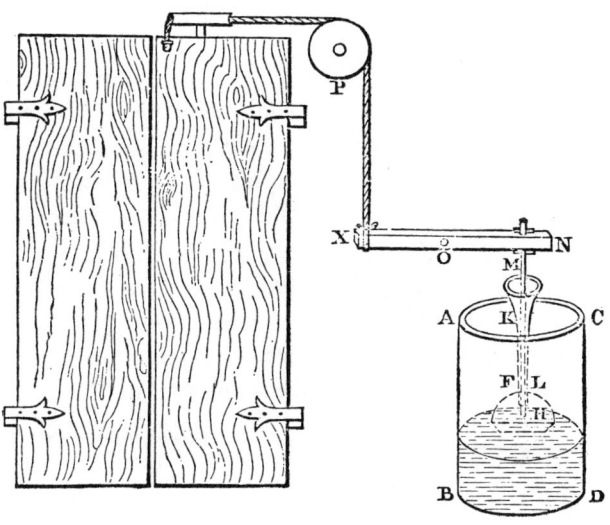

the "mechanical" explanations of the Milesian engineers and their successors, the proponents of the atomic theory. Not only did his proposed compromise force him to synthesize the very notions of a mechanical model and a pure mathematical set of equations, but it also forced a choice between two opposed explanations of cosmic motion. Plato and the formalist tradition interpreted the regularity of cosmic motion as a sign of life and intelligence on the part of planets and stars. The atomist and other materialists, on the other hand, argued that the mechanical regularity of the heavens showed that they are an *unintelligent mechanism,* and that transfer of momentum—going back infinitely far, since "the atoms have always been in motion"—was the correct type of causal explanation.

Aristotle accepted, as his predecessors had, the notion that the heavens were absolutely unchanging. As evidence for this, he remarks that Babylonian observation records "over a period of six million years" had revealed no change. Whether this report was mere prevarication on the part of the Babylonian scientists, or poor scholarship by Greek travelers, it served as unquestioned fact for Greek cosmologists and model builders.

The deadlock seemed to Aristotle to call for the invention of a wholly new science of "celestial mechanics." The mathematicians of the Academy, had, as we have seen, designed a geometric model of concentric spheres, each with its own speed and direction of rotation. Older models had shown that a two-component mathematical theory could be represented and duplicated mechanically—given at this stage an external source of motive power.

Aristotle started his new science with the assumption that laws of motion must be modified for celestial motions. All terrestrial motions in Aristotle's scheme come to a stop unless "energy" is added. For *natural* terrestrial motions, Aristotle envisaged the "center of the world" as the earth, surrounded by concentric layers of water, air, and fire, each with its own specific gravity. Each element therefore moves naturally up or down toward its "proper place", where it comes to rest.

Celestial motions, on the other hand, are naturally circular, not linear. Being circular, they never can reach a "proper place" where they come to rest. In effect, the heavens are a perpetual-motion machine. At the start of his astronomy, therefore, Aristotle introduced the hypothesis that the stars (and spheres which move them) are made of a "fifth element", *aither,* chemically different from the terrestrial four elements. *Aither* moves naturally in a circle, and does so without encountering resistance and without inertia. Thus models designed for "celestial mechanics" will themselves fall somewhere between the pure ideal abstractions of mathematics and the simple mechanical analogue made of terrestrial "hardware".

The nested crystal-like spheres that Aristotle envisaged have two other properties, in addition to their circular motions. Each one, as it moves, carries around all the others within it. Thus the daily revolution of the outermost "first heaven" carries the whole inner system around with itself.

At this point, the difficulty of suitable model design must have become clear. Eudoxus' mathematics had worked out the number of component rotating motions needed to duplicate the apparent motions *separately* for sun, moon, planets, and stars. But since these motions are those of nested concentric spheres, they are not mechanically independent even if a postulated frictionless matter makes up the mechanism. Aristotle's innovation was to add "counteracting spheres," moving in such a way that the outer sphere of each planet, and that of the sun and that of the moon, were insulated from the effect of the complex motions further out in the cosmos, except for the daily revolution of the "outermost" sphere carrying the fixed stars. The upshot was a proposed "model" in which the total number of spheres was "either 47 or 55."

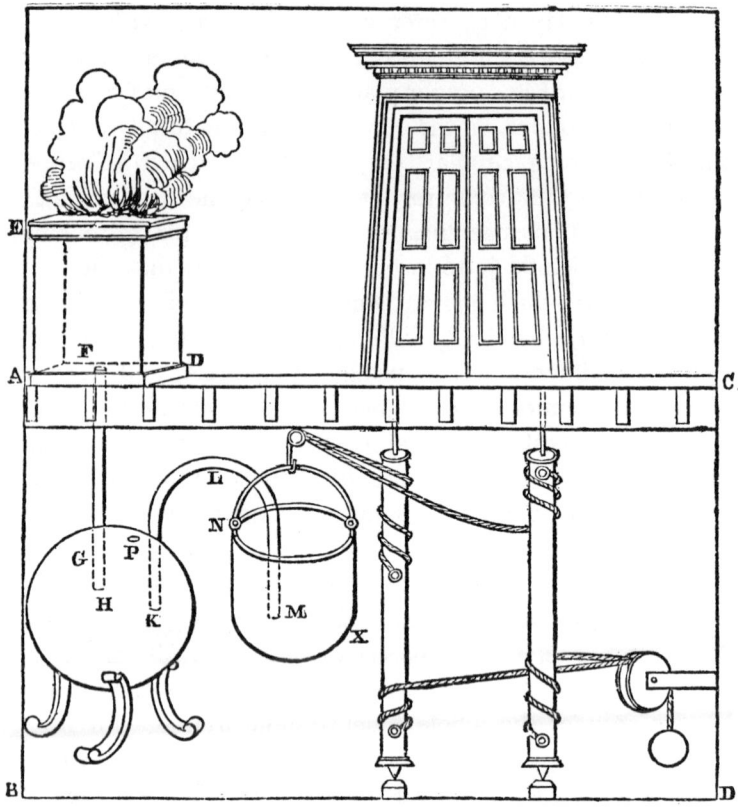

Self-opening temple doors were almost certainly originally operated by concealed assistants who pulled the cords wrapped around the doorposts below floor level. Heron's device is a more sophisticated way to make doors open automatically when a fire is kindled on the altar. The heat of the fire caused water in the hollow sphere to overflow in the bucket. The bucket dropped, pulling the cords along with it.

Did Aristotle think his model could be built? And, regardless of what he thought, could it be? My friend Professor Ronald Levinson proposed this question to me years ago, when we were corresponding about Plato's astronomy. I think we can come close to building Aristotle's cosmic model today, which shows that it does not involve the inner mechanical *impossibilities* that many scholars have attributed to it.

The design I worked out would operate electrically. Each sphere has a small motor (possibly with reduction gearing) at its pivot, driving its shaft. The spheres are not in contact, but around each is a copper band, with an inner and outer pair of brushes that transmit and pick up electricity. Colored spots or lights indicate the sun, moon, planets, and stars. The device as a whole turns on a pivot for the daily revolution.

But though it is possible to design such a modern planetary path generator, the sheer complexity of "47 or 55" of these spheres would push even our modern technology to its limit. Actually, Aristotle himself seems to have miscalculated the number of "counteracting spheres" his model would need, as a two-dimensional projection by my colleague Professor Norwood R. Hanson has shown.

Where was cosmological model building to go after this ultracomplex blueprint by Aristotle? In one tradition, it turned to computation, without geometry. In another, it went on devising simple but self-moving models of a strictly mechanical kind, avoiding the need for "spiritual force" or "cosmic mind" that Aristotle had postulated. In the Hellenistic period, the Epicurean school revived a counterpart of the old "vortex" model (with streams of particles rather than continuous matter), thus continuing the mechanical aspect of Aristotle's machine. The Stoics, on the other hand, were impressed by the idea of a divine "spiritual force," and we will find one later record of one of their models showing how such a force supports the earth at the universe's center.

Of all his syntheses of ideas from the formalist-mathematical and the materialist-mechanical traditions, Aristotle's astronomical theory and his cosmic model, despite its aesthetic appeal, must be judged the most complex and the least successful.

CHAPTER 5

MECHANICAL MARVELS

THE SIXTH AND FIFTH CENTURIES B.C. marked the period of the brilliant and flamboyant rise of Athens to its leading role as Greek cultural and business capital. Here in the markets, the schools of philosophy, the courts, and the puppet theaters, all four of the main themes we are tracing—automata, cosmological models, the mechanical insurance of honesty, and (to a much less degree) use of nonanimate power sources—undergo important developments. The everyday world of the ancient Athenians had running through its texture an important strand of interest in mechanism and ingenuity of mechanical design. Two things are particularly interesting about this. The first is the extent to which machines could be designed to do things which are like the "purposive" behavior of a human agent. Puppets and marionettes are mechanisms of this type—"amazing things" that fascinated the Greeks because of their likenesses to human nature and reality. The second is the role that mechanisms played in almost every part of Athenian life, with such devices as specialized kitchenware, children's toys, adult games, shopping transactions, elections, and legal processes.

But before we turn to the puppets and the philosophical questions they raised, let us explore quickly the more mundane devices found in the Athenian home.

A favorite of mine is one of the oldest: a wine cooler

MECHANICAL MARVELS 47

(*psykter*) from the sixth century B.C. This was a wine jar of ordinary shape, but with coils of clay tubing running about inside, connecting with openings on the outside of the jar. The design is the same as that of the water-heating coils that ran through coal furnaces in my younger days, but the purpose was, of course, the exact opposite. Immersed in cold water, this would circulate through the coils and chill, without diluting, the wine.

Another device of this time is known only from a reference in the philosophic writings of Aristotle—namely, a set of hollow toy frogs. I wish our archaeologists had found some. In a solemn explanation of the way that air rises as bubbles to the surface of water, Aristotle remembers these little toys and uses them as illustration. Filled with different quantities of salt, and placed under water, these little batrachia would pop to the surface when the salt dissolved. Proper timing could probably produce the effect of a choric dance. And, human nature having stayed much the same, I am sure that occasionally these found their way into the family bowl of wine, leading to some hard feelings.

A much homelier device that impressed me as "gadgety" was an improved design of sausage grill. When the ancient Greeks cooked meat—which, by the fourth century, was relatively infrequently—they grilled it on skewers or laid it in strips on a standard ceramic grill over a charcoal fire. As every home chef knows, meat on skewers or grills tends to roll, resulting in a biting of the dust for sausage and sausage consumer alike. Unsung geniuses of ancient Athens had fixed all that: one new grill design of the fourth century had ceramic lugs sticking up to hold the skewer in place, while another had small spikes along the grill strips, to keep roasting sausages from rolling as the barbecue went on.

For some reason, the kitchen has always been a fertile field for gadgetry. A visit to Tom's Market, our local grocery store, which also carries a line of kitchenware, offers plenty of con-

Heat from the fire, blown through bent tubes, caused the platform to revolve. Heron also used the same principle for his steam engine.

temporary examples. Among them, picking at random, are: the wire egg slicer; the special grater for cutting uniform thin slices of American cheese; the carrot scraper; the apple corer; a runcible spoon, with a cutting edge, facilitating the serving of cranberry jelly; a set of two-pronged yellow plastic skewers, to assist with neater eating of corn on the cob; and about half a hundred more. Hope, apparently, has been continual, from ancient Athens to modern times, that the tedious side of food preparation could be circumvented mechanically, leaving the cook free to concentrate on his or her humanistic artistry. When the cooks in question were slaves or servants, and cookery was not very good (as it was not in ancient Greece and Rome), gadgets promising improvement in quality of food were, I suppose, in particular demand. But even there the cook seems to have had a certain artistic license and persuasive rhetoric that created a market for new ovens, pans, and strainers designed to save labor. A fine selection of pictures is in the American School of Classical Studies, Excavations of the Athenian Agora Picture Books, #1, *Pots and Pans of Classical Athens,* prepared by Brian A. Sparkes and Lucy Talcott (3rd printing, 1964).

Novelty pitchers of varied design continued in production. Two from the Athenian Agora are an athlete's oil flask in the shape of a kneeling boy racer, and a more austere milk or water jar in the shape of a hollow woman's head. Lamps, too, were beginning to try for new effects. Souvenir statues were mass produced from molds as early as the fifth century B.C. As time went on, this miniature sculpture became more fantastic and cute. I can imagine clay rabbits nibbling leaves, or toy turtles of baked clay, being bought for the kids by a traveling father.

Out of respect for Dr. Spock and others, I hesitate to include the classical potty-chair here as a mere gadget. But it obviously should be included in any history of human ingenuity, so I mention in passing that one of these was found intact in the excavations of the ancient Agora. Since its design has become a standard Western model, even down to the pictures to entertain the impatient seated one, I merely indicate its existence and early date.

There were also all sorts of games and toys. Large and small balls, kites, tops, dolls, doll beds, rattles, toy drums, dice, and checkerboards with pebble pieces were among them.

Vase paintings and descriptions also record a popular indoor sport, *kottabos*, which was in fashion. We can think of this as an ancestor of our modern game of darts, though in this ancient game, splashes of wine, not darts, were thrown at a target. The target was usually a floating copper bowl. A two-dimensional figure of Hermes stood on its rim in the deluxe *kottabos* sets. Points were scored when Hermes was hit directly—with a ping—for a rebound from Hermes into the bowl, extra for sinking the floating bowl. Some houses, we are told, had special round gamerooms for this diversion (round to place players, who reclined on couches, all an equal distance from the target). There is a vase painting of a much more elaborate set, which reminds me of the "80 free game" circuit—lights, flags, and bells—in my old pinball machine.

So the parade goes on, from sausage grills to pull toys. All this, however, as far as ancient Athens is concerned, is a mere tributary stream to the main flow of technology and gadgetry.

Then there were the classical ancestors of our modern flood of windup and electric toys. These were a subdivision of what the Greeks called "amazing things", puppets and marionettes. There were marionettes, like our own, operated by control strings. There were "puppets"—probably translucent two-dimensional ones—which projected colored figures on a wall. And there were also the self-moving windup type. Their behavior was lifelike enough to seem spooky even to scientists and philosophers. A stick was fitted through a center loop in a leather thong; this was wound up, and when it was released, the stick knocked against "pegs" as it spun. The pegs were connected by levers to the head, arms, and legs of the puppet figure. Some of our modern psychological notions about the human nervous system were anticipated, in a crude form, by a fourth-century author, who wrote that "people respond to the impressions that strike their senses like self-moving puppets." In a later century we will find this marionette tradition threatening human actors with technological unemployment in a grand design for an "automated theater".

Athenian mechanical marvels were dominated by "amazing things"—puppets, marionettes, and other automata. Apparently these were of several types. There are no exact descriptions or intact specimens of their "works", but we do have many references, including Aristotle's list of "wonders" that inspire philosophy. In the history of gadgetry there is, I believe, a law of conservation of design. Modifications take place slowly, and usually only by one new improvement at a time. Therefore, it may be possible to find out more about the actual mechanisms by looking at their descendants, the mechanical theaters, for which details are given.

In his famous "Myth of the Cave", Plato compares the knowledge of uneducated political operators to the "political scientist" who understands human nature and conduct. Plato's myth illustrates the difference between the uncritical "guess" and real "knowledge". The former is the kind of knowledge of realities a spectator would gain from watching shadows of puppets which he mistook for real things. Imagine an audience imprisoned in a cave. In front of them is a

Greek mechanics worked out fairly sophisticated designs for dolls or toys on wheels that would stop, then start again at right angles. In this diagram, seen from above, a larger set of wheels is held in a raised position. When they are lowered, the machine rests on them, and the result is a right-angle turn.

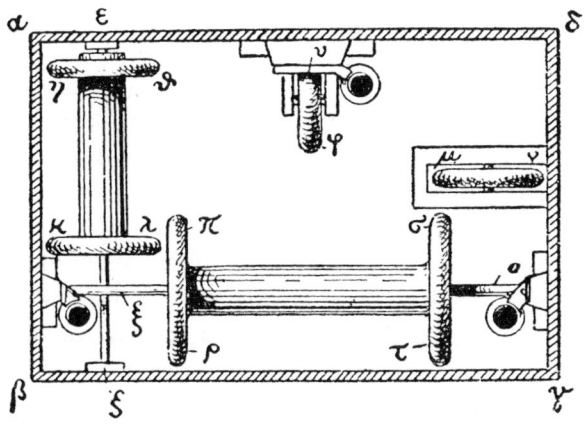

wall and behind them a parapet and a fire. Hidden behind the parapet, puppeteers carry their dolls in front of the fire so that shadows are projected on the wall, and the voices of the operators are echoed from it as well. Would not an audience that had never left this prison think of these shadows as the only "realities" there were? Might they not even set up guessing contests, giving prizes to those who are best in anticipating what shadow will come next? Strange prisoners, but "like ourselves" when we begin our search for knowledge on a level of mythology, guesswork, and poetry. Education can, and should, lead us from this level—out of the cave, past the marionettes—to a world of sunlight where we see the originals, from which the puppets are copied, not by dim firelight but in the light of day.

In this story, Plato combines two different kinds of "puppet". In the Near East today there are traditional puppets which are two-dimensional and translucent, colored outlines as they pass before the light backstage. Their shadowy appearance suggests the "shadows on the cave wall", and it would be my guess that Plato had seen some of this design. On the other hand, the actual puppets carried by are *not* these. Those are "some of wood, others of stone, others made of all other materials." We want them to contrast with their shadows in the way that puppets contrast with the originals they copy, to carry out the scheme of appearance—reality. There must be something "more real" about the wooden, stone, and other figures than there is about their projections. This might be achieved by making them three-dimensional or by making them move themselves in a way that "causes" the motions of the shadows, or perhaps both.

Two centuries later a mechanic, Ctesibius, set out to design an automatic theater. Four centuries after that, Heron of Alexandria published plans for such a theater. In his plans Heron indicated a number of new improvements (a thunder effect, a fire effect, and a lightning bolt) over Ctesibius' design. The inference seems to be that the *other* features of Heron's theater were traditional, and in most cases like those Ctesibius had used. It is certain, for instance, that the same five-scene play Heron's theater presented was the one Ctesibius first chose for his theater. In the Ctesibius-Heron theater the "actors" were silhouetted against varying backdrops, running across the stage on various parallel tracks. A gravity-driven "motor" and set of wheels replaced Plato's concealed showmen to carry the puppets along. Concealed internal "works" made the puppets move and act out the play; yet the visual effect was two-dimensional. We know from another Platonic passage that puppets were often "worked by strings." There is general agreement that these strings were not operated from above, as with our modern marionettes, but from below.

This description permits us to reconstruct the idea Plato had in mind. His puppets were not simply to be carried woodenly. In contrast to their shadows, they appeared to be "alive," owing to their string-operated motions. This power of self-motion is, of course, only apparent. Plato echoed older tradition in taking "self-moving" to be an essential and exclusive property of things that have a "soul". And the hypothetical prisoner in the cave would recognize this, once he had gone out into the world of nature. There he would see the difference between the automata and a living, adapting organism.

However, neither the silhouette casting shadows nor the marionette operated by strings from below was the truly "amazing" automaton in ancient Athens. These were, rather, the dolls that exhibited lifelike behavior "when stick [or "peg"], unwinding, struck the *neura*." *Neura*—"strings" or "cords"—is the word also used in anatomy for "nerves" or "tendons", another interesting indication of the interaction of the ideas of "animate" and "automatic" behavior. Here, again, Heron's designs are helpful. Since he does not claim any part

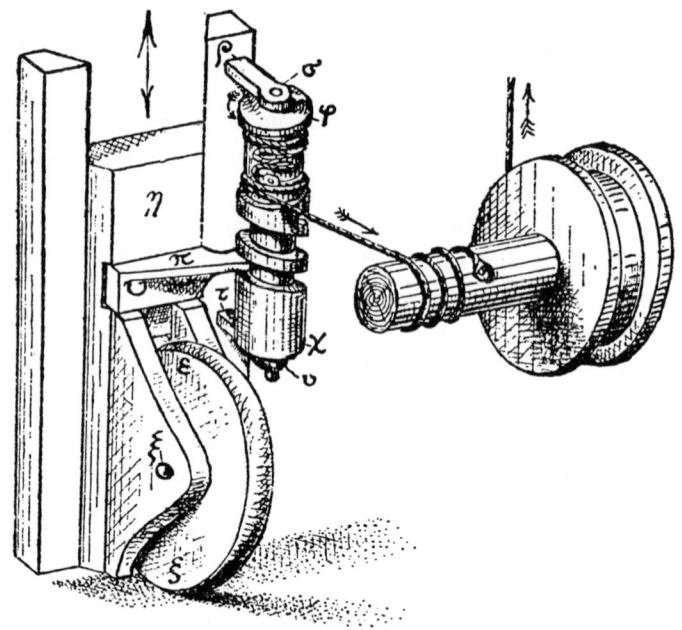

Detail for right-angle motion of automaton. As the screw turns it lowers the plate. The plate is fastened to a sliding panel. A wheel (one of the larger set that are lowered for the turn effect) attached to the panel is forced down with it. There is no evidence that this design ever really worked.

of them for his own, they presumably go back to Ctesibius. He, in turn, inherited them from an older tradition. The first scene of the play *Nauplius,* in Ctesibius' (and Heron's) theater, consists of nymphs moving across stage against the proper backdrop as they build a ship. They are seen hammering, sawing, and boring holes, "just as though they were real." The mechanism producing this effect is a wheel with projecting pegs around the rim. In Heron's model the wheel is turned by a falling weight. The pegs strike the end of a bar that moves the arms of the shipbuilders. The "realistic" hammering effect is achieved as the arm rises and falls. If we can substitute lighter strings ("tendons") for the relatively heavy bar, using a shaped "peg" or "pegs" for the projections from Heron's control wheel, we have something that exactly matches the references to the classical self-moving automaton. The "windup" power was probably supplied by a twisted leather thong through which the "peg" fitted.

The popularity of the puppet coincided in ancient Athens with a crisis in Greek psychology. The central character in this drama was Socrates, whose insistent questioning into the nature of a human "self" changed the course of Greek philosophy and led to his own trial and execution. As a young man Socrates was fascinated by the "scientific" ideas which had just come to the attention of Athens. Two problems that he was particularly concerned with were "whether the earth is cylindrical or round, and why it rests in the center of the universe" and "whether we think with the air in us, or the fire . . . or perhaps with the brain." The second of these problems reflects the state of scientific psychology at the time. Although popular religion had a "soul" which survived death, this was a sort of shadowy body; and science and religion had gone separate ways. The scientists and physicians were moving in the direction of treating the "soul" as the physical organ that thought and moved the body, and trying, by dissection and experiment, to localize it. According to the atomic theory, the "soul" was a mechanism "made of very fine, round atoms." Greek literature and common sense tended to identify the "self" with the body it inhabited to a surprising degree, as Snell has demonstrated in his study *The Discovery of the Mind*. Against all this, Socrates suddenly realized that physical models simply did not match human behavior. It was not that the "soul" merely registered the adventures of its body; rather, it directed them. For example, the ideals that men held could cause them to behave in ways contrary to the comfort and even survival of their bodies. Further, the process of learning, as one found it in pure mathematics, was something more than a mere accumulation of fact. Exactly what a self was, Socrates did not know; but that it was not a physical organ he was sure. He was equally sure that no one would know what was intrinsically good for man until this nature was found out, and that the politicians should be provoked into inquiring. At a crisis in Athenian history when the politicians wanted to remain unquestioned, Socrates' constant challenge seemed dangerous. He was accused of impiety, and was executed in 399 B.C.

The account of this is given by Socrates' greatest student and admirer, Plato, in his dialogues *Euthyphro, Apology, Crito, Phaedo*. The analogy of marionettes does not figure in this report. But Socrates is shown rejecting the idea that the soul is simply an "epiphenomenon," an awareness of happenings which the body causes. The model he is attacking would reduce explanation of human behavior to "tense sinews" causing motion, or to "strings under tension" producing a harmony. And these indicate that puppets have had their suggestive effect on physiology and psychology.

With Plato the problem is actually stated in terms of the analogy of marionette and man. Plato explores the possibility that since the soul is a nonphysical thing it may be immortal. One of his arguments is that the soul has the peculiar property of self-motion, whereas physical things can be moved only by transmitted motion. If, then, man is compared to a "puppet of the gods", the comparison fails because man is able to initiate his own actions. He is a puppet that is not wound up, but pulls its own strings and writes its own script for the play as it goes along.

Plato's student Aristotle tried to bring together the insights of his teacher and those of the atomic theory. The soul, he concludes after examination, is neither an organ of the body nor an entity separate and distinct from the body. It is, rather, the way of functioning of a properly designed organism. Thus physical organization is a necessary condition for the presence and activity of a self or soul. Aristotle tries to have *both* physical and mental causality in his scheme. Again the marionettes are in the background of Aristotle's thought; they are not "alive", but we can learn about living things from their duplication of animal and human behavior.

After Aristotle the debate over man and soul was continued in Hellenistic philosophy, where improved automata made the mechanistic view even more convincing. The final act of the drama, for the present chapter, is found in *De motu*, a work by a student of Aristotle's. Starting with Aristotle's observations that "organs" and "organisms" show an adaptive

Axle-winding patterns show designs for forward motion (*bottom figure*) and forward and reverse motions (*top two figures*).

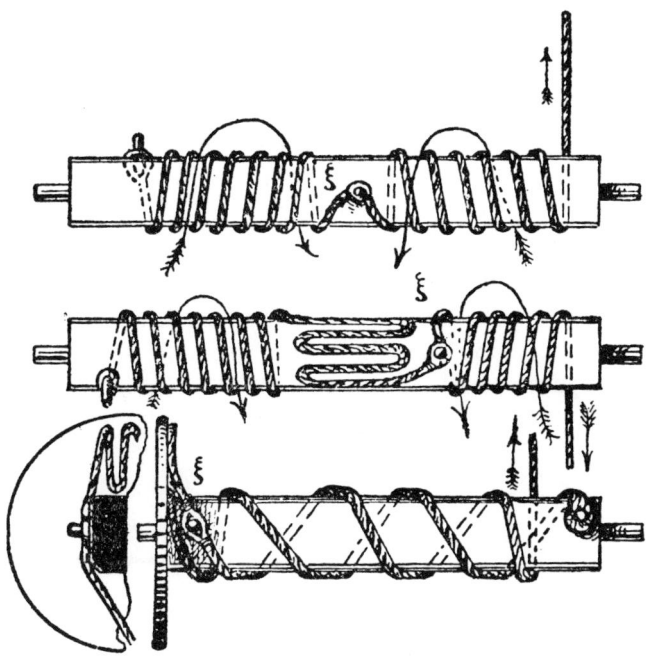

and flexible kind of response that ordinary string-and-lever "mechanisms" do not, the author goes on to a new theory. Suppose the soul is a central organ filled with fluid, shaped like the body, and connected by ducts. Then our bodily contacts compress this by transmitting pressure. The organ, being elastic, springs back to its normal shape. Some fluid is forced out into other channels, its motion is amplified, and we get an appearance of adaptive response. This author is explicit in his conclusion, that in this way behavior of human beings is like that of the marionettes. In effect we seem to have an anticipation of the modern psychological concept of the stimulus-response reflex arc, an adaptation of Aristotle which goes further toward the mechanical and away from the Platonic side of the question.

Without mechanisms that could be designed to imitate living things, this chain of events in the history of ideas might have been a quite different one. For example, in the early stage of panpsychism, nothing short of "wonder at the puppets" would have been needed to give any credibility to the atomic theory. And without that theory's suggestion that man is really a machine, speculative philosophy might never have been so concerned to define the difference between what is vital and what is merely mechanical.

Philosophers today are facing a similar question. The development of computers that "think" makes us wonder what limits there are to the tasks a computer can perform, and what we mean by "thought." "Can machines think?" is the modern counterpart to the ancient Greek question, "Are some machines alive?"

CHAPTER 6

THE DEMOCRATIC LOTTERY

In its experiments with democracy, ancient Athens interpreted "equal opportunity to hold office" very literally. Selection for all public jobs, except those of military officers and treasurers, was by lottery. An account is preserved of the constitutional provisions for electing an Assembly, deciding which of its ten component tribes would assume direction for each month, and selecting juries from a large panel. A look at the table of offices filled by lot goes a long way toward clarifying the doubts about "democracy" expressed by the Greek philosophers Socrates, Plato, and Aristotle. Instead of criticizing these philosophers for their "aristocratic, leisure-class" outlook, as has usually been done, scholars would have done well to imagine a modern city with its offices filled in the grand lottery style! If aldermen, park commissioners, police chiefs were picked by pure chance, this could turn a city overnight into a monument to shambling ineptitude unless some goddess of luck took a helpful hand in matching duties with qualifications.

Athens offers a remarkable case study in the interaction of mechanical devices and ethical concepts, and of the philosophic questions which this interaction suggested. In addition to the concept of "equal eligibility for office" was the notion of "fair trial by jury" in the same period. "Equal eligibility" was interpreted as "equal opportunity" and then operationally defined as "equal probability" through the institution of lottery selection of officials. "Impartiality" in a similar way is defined as "selection of jurors at random," again with the lottery giving an operational procedure. As a safeguard against unfair sampling, the juries were enormous by our standards: two hundred, five hundred, one thousand were impaneled for varying cases. A look at the mechanical side of this affair may be helpful both in clarifying our understanding of the way in which classical "democracy" differed from our own, and for some of the criticism of "democracy" by political theorists of the period.

A second major area of interaction between concept and mechanism was in the field of business. From a strange mixture of barter, bargaining, gift exchanging on a commercial basis, and general piracy, standardized weights and measures, including trademarks and coinage, brought some law and order. With it came a definition of equality in a new sphere: the idea of fair measure, the same regardless of the customer's gullibility or his apparent ability to pay. Also, though more gradually, a notion of standards of quality began to appear.

These two developments suggested an interesting philosophic question. Since standards were arbitrary, and differed between Athens and Lydia or Persia, one group of social theorists argued that this proved that all social values are matters of arbitrary local convention. Other thinkers believed that the objectivity achieved by standards of election or trade was taking advantage of "laws of nature" to stabilize human behavior, and that "human nature" was the same for different places. Though the Lydians had a different scale from the Greeks, still they too had standard measures and were no more eager to be shortchanged!

In the present chapter we will look first at the elections and jury selection by lottery, then at the regulation of business, and finally at the ultimate in public regulation—the introduction of official time measured mechanically by a clock.

Three or four different procedures were used in the ancient lottery. The greatest need for precautions was in jury selection. It seems to have been axiomatic that if anyone knew in advance which jurors would hear his case, intimidation and bribery would result. And it is in this field that we find the ancient counterpart of our modern voting machines, as mechanical ingenuity was harnessed to the automation of honesty. The constitution itself specified use of certain foolproof devices, to ensure an honest day in court.

The oldest lottery system consisted simply in simultaneous drawing from two boxes: the one held name tickets for eligible citizens, the other a mixture of black beans and white. If a white bean was drawn, the ticket owner was selected; rejected if a black one appeared. Later a second device was used, apparently to determine the order in which the ten "tribes" would hold chairmanship of the council, each for one month of the year. These were ceramic tokens, cut in half in jagged patterns, bearing names of tribes; apparently one half would be drawn, and the holder of the matching piece would be selected. A set from the fifth century has been found. By mid-fourth century, an ingenious lottery machine, the *kleroterion,* had been devised; twenty were used to impanel a jury for one day in court.

With little ingenuity the ordinary lottery could be prearranged, a fact that was generally recognized. Plato, in *Republic* v, speculating on ways to regulate marriage for the general good, decided that the best men and women, and the worst, should be paired together as parents at the marriage festivals. "This we will arrange by an ingenious system of lots, which the rulers will contrive to fall out rightly. In this way, each man will blame his fortune on the lottery, not on the rulers themselves. . . ." The general principle of this well-intentioned rigging probably presupposes two sets of name tickets, for men and women, drawn from two boxes. The system can be controlled by having matching orders in the two sets, so that good and bad always match. Various elaborations—cutting stacks of tickets like decks of cards, pinpoint marking that can be felt by a drawer of lots, hoppers or boxes that will shake tickets without disordering them—can make this simple principle less obvious in operation. The point is that, at least where small numbers were involved, everyone recognized that the bean-and-ticket type of lottery was far from tamper-proof.

A second important application of gadgetry to ensure honesty had to do with the secret ballot. Jurymen, for their protection, needed to have their votes kept secret. And the gen-

Heron's self-trimming lamp probably never worked. In this design, the sinking weight is supposed to rotate the gearing mechanism, which pushes the rod *E* forward, forcing out the wick at *C*.

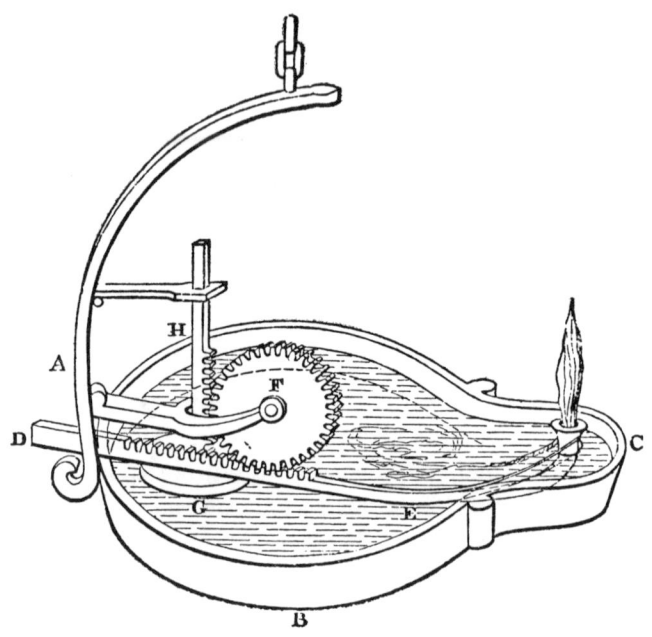

eral public, where a large number of votes by pebbles or beans was cast—for instance, in deciding whether the archons and judges were qualified for office—needed some enforcement of a "one man, one vote" procedure. The final results were wheel-shaped ballots and a special voting box which still commend themselves as political applications of mechanical ingenuity.

We will look first at the impressive list of city officials selected annually by lot; then at the explicit constitutional provisions for jury selection and secret voting. The particular apparatus to be used is specified only in this latter case; but it is a safe assumption that, as jury lotteries became more tamper-proof, similar procedures found their way into the general election procedure.

A quick look into Aristotle's *Constitution of Athens* gives a fair impression of the extent to which Athenian administration depended on Tyche, the goddess of luck.

43. So much for the rules about the enrollment of citizens and about the young men of military age. All the officials for the ordinary administration are chosen by lot, with the exception of the Treasurer of the military funds, the Treasurers of the theater fund, and the Superintendent of the water supply. These latter officials are elected by vote . . . .

The Council is selected by lot and consists of five hundred members, fifty from each tribe. Each of the tribes holds the Presidency in turn, the sequence being determined by lot; the four on which the lot falls first hold the Presidency for thirty-six days each, and the remaining six for thirty-five days each . . . .

44. The Prytanes have a chairman who is chosen by lot. He holds the chairmanship for one day and one night . . . .

47. The Council also collaborates with the other magistracies and the greater part of the administration. There

are in the first place the Treasurers of Athena, chosen by lot, one from each tribe . . . .

Secondly, there are the ten Poletae, chosen by lot, one from each tribe. They farm out all the public contracts . . . .

48. There are ten Receivers, collectors of public revenues, one chosen by lot from each tribe . . . .

The Council selects also by lot ten Accountants from its own members. These men have to check the accounts of the magistrates during each prytany.

They also select by lot one Examiner from each tribe, and two associate examiners for each of the Examiners. (These make preliminary decisions in charges involving legal cases.) [1]

Aristotle's list of officials selected by lot goes on: ten commissioners for maintenance of sanctuaries, eleven wardens of the state prison, fifty-five supervisors of trade and markets, five introducers who draw up court agenda for cases to be tried within the month, a board of forty judges, and so on.

The system included elaborate checks and balances; for example, "ten Auditors and ten Assistant Auditors, to whom all those who have held public office must render account." The nine archons, the highest officials, are examined before the council of five hundred. "If nobody brings a charge (against the candidate), he (the examiner) puts the matter to the vote at once. In former times, one judge only used to cast his ballot into the voting urn (in such unchallenged cases) but now all the judges have to vote about the candidates so that if a dishonest man has got rid of his potential accusers, it is still in the power of the judges to reject him."

Perhaps the ten most unenviable jobs were those of the market inspectors. They were charged with enforcing a max-

[1] Aristotle, *The Constitution of Athens and Related Texts*, ed. and trans. by K. von Fritz and E. Kapp (New York, Hafner Publishing Co., 1950).

imum two-drachma-per-night rate for flute girls; and, in cases of dispute over a flute girl for the night, they decided which claimant to her services received them by holding a lottery!

Turning now to jury trial, we come upon the masterful *kleroterion*—the foolproof lottery machine. A stone backboard contained rows of slots; metal tickets of eligible candidates were inserted, column by column. Fixed to the left side of this stone backboard was a large hopper, into which balls were poured: white ones equal to the number of candidates or groups of candidates to be selected, black ones to bring the total up to the total number of rows. Apparently, the balls in the hopper were then stirred by hand, or mixed and poured in, in some random way. Unfortunately the constitution does not tell us this detail. Obviously it would be better if the hopper itself were movable, mixing the balls mechanically. But finds to date do not completely decide the question of the method of attachment to the backboard, and the proposed reconstructions do not provide for a movable hopper-tube unit. From the hopper, attached to the left side of the machine, a tube admitting one ball at a time ran down, with a gate or crank at the junction of tube and hopper. The balls would be extracted one by one, and each row cancelled or appointed depending on whether the ball drawn was white or black.

Citizens eligible for jury duty assembled by tribe, each bearing a ticket with his name and a letter indicating to which of ten subordinate jury groups he belonged. (Later, these subordinate groups were varied each time by random drawing of acorns, lettered A to K.) The name tickets were then placed in the slots of the machine, each letter group in its own column. The attached bronze hopper was filled with black and white balls. A ball was drawn for each horizontal row; a white ball selected that entire row for jury duty, and those tickets were set aside. Next, in the earlier phase, each juror drew a bronze ball marked with a letter from another

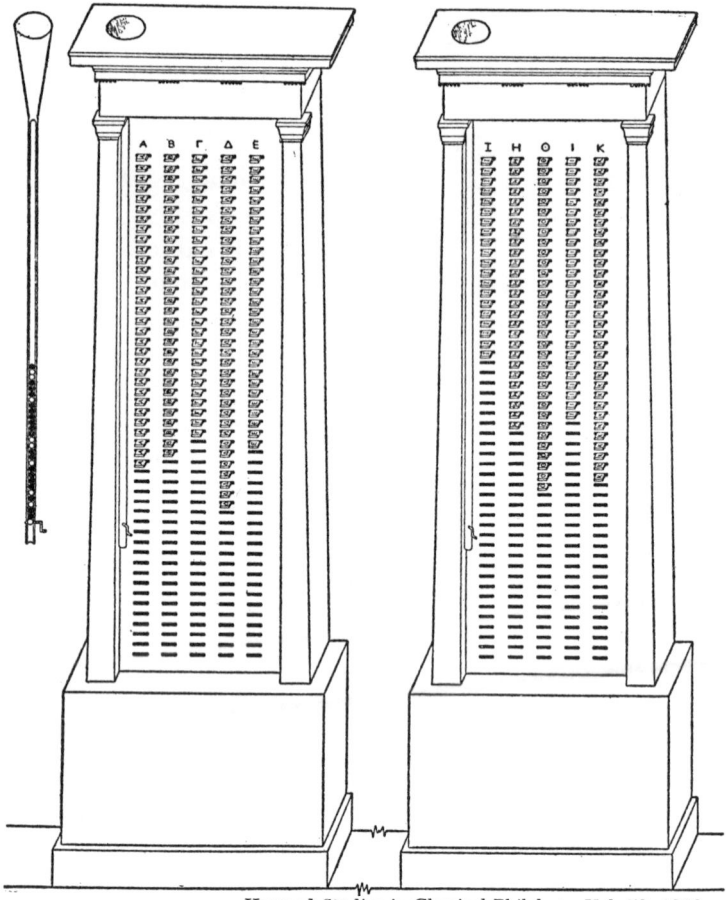

Harvard Studies in Classical Philology, Vol. 50, 1939

Equal opportunity to hold public office in Athenian democracy was interpreted as equal probability that each citizen would be selected. Juries, and probably other executive officials as well, were chosen by an ingenious lottery machine, the *kleroterion*, devised in the fourth century B.C.

box. This letter indicated the court chamber to which he was assigned; his ticket was sent there, and he could only receive payment in *that* court. But by Aristotle's time, some new precautions had been added. A lottery selection decided the color of a staff given to each juror. Then use of dice with faces of six different colors decided which color staff was to go to which chamber of the court. Wardens admitted only the right color, collecting the staffs at the door. The juror was given a special "token" entitling him to payment; this to guard against unauthorized interlopers joining the pay line at the conclusion of the day's cases. (The juries were very large —two hundred for petty larceny cases, for example—so that personnel control offered many problems and many chances for beating the system.) And now, other dice were thrown to determine which judge and which case should go to which chamber of the court. A small model of the court building was used to keep track of these assignments of judges, cases, and jurors.

Pleading time was limited to fractions of a standard day, its length varying with the type of case. Upon arrival in the court, a further lottery was held to select one of the jurymen as keeper of the water-clock. Later, the case having been heard, it was time to vote. Two ballots (*psephoi*, "pebbles," a name carried over from an earlier state of innocence) were given each juror. They were metal wheels with shafts extending slightly on either side, identical except that one shaft was solid—this was the acquittal ballot—and the other shaft hollow—a vote for conviction. These were to be picked up by each juror "from a lamp-stand", a phrase still not entirely understood; but it is clear that no one could tell which ballot was put into the voting urn, and which into a discard box. Everyone could tell, however, that only one ballot at a time went into the urn that counted; for this had a solid top, with a slot cut in the shape of the ballot wheels, that admitted one and only one ballot at a time!

After the vote, a special counting-board was used to tally the ballots cast; the jurors traded their tokens for pay, and everyone went home.

Not only did Athens try to channel juries and elections into mechanized honesty, but there also was a veritable passion for business regulation. The general attitude toward the market seems to have been that one could expect to be cheated unless the government stepped in. How true this was, I don't know; but I do recall that one author proposed regulation of hotels, to prevent the practice of "holding guests for ransom": and this practice was probably matched in other transactions.

In addition to the ten market inspectors in charge of flute girls, the election by lot chose ten supervisors of weights and measures, and there was a paid force of market police. A special building, the Metroon, served as the local bureau of standards. Excavations have brought to light a formidable set of official weights and measures, plus some seals of the inspectors. Apparently, in response to public demand, standardization extended more and more widely. Finds include official dry and fluid measures of the fourth century, an official basket for the measuring of nuts in the third, and a standard roof tile mounted in a stone frame that was put up in the market in the first century B.C. There may well have been one earlier, but to date it hasn't been found.

This complex of weights and measures probably was important for the history of ideas in the West. At any rate, it must have gone far to reinforce the idea of the Greek Sophists, in the fifth and fourth centuries B.C., that society rests on arbitrary standards set up by convention. It certainly belongs here as a part of the theme of the mechanization of honesty, though the techniques used for standard measure are so straightforward that they probably do not count much in a survey of "gadgetry."

The possible exception is the Agora "town clock" of about 350 B.C. This was a large water-clock, operated by a float which sank as a tiny outlet at the bottom emptied a well. The float, as it sank, turned a shaft to which either a globe with a picture of the sun, or a single large pointing hand, was prob-

THE DEMOCRATIC LOTTERY 69

ably attached. The amount of wear on the small steps leading down into the well suggests that this clock had to be "rewound" by some official every day. I find it fascinating to realize that ancient Athens was already so much like a modern city that it was running on clock time in the fourth century B.C.

Perhaps the most far-reaching development in connection with the interaction between objective, mechanical standards and patterns of human behavior was this introduction of pub-

This "town clock" of *ca.* 350 B.C. was operated by water which emptied through the drainpipe in the foreground. A float which sank as the water level lowered in the well turned a shaft to which the clock's mechanism was attached. Reported heavy wear on the small steps leading to the well suggests a daily "rewinding" or replenishing of the water supply.

American School of Classical Studies at Athens

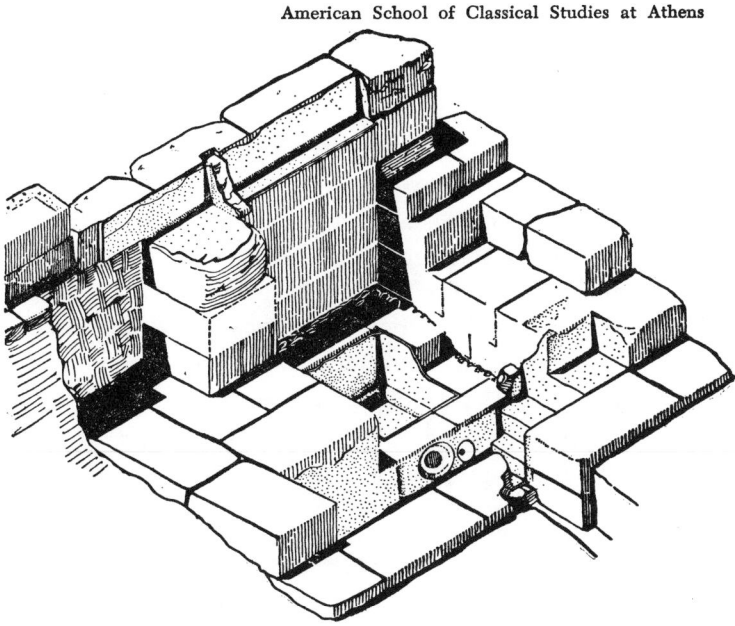

lic clocks and uniform mechanical time. Ordinarily we think of the era of the clock as starting in the fourteenth century, with the magnificent astronomical clock of Da Dondi, built for the cathedral of Strasbourg. We sometimes look back wistfully to the world of ancient Greece as one of leisure, of "life time". But in fact, by about 360 B.C. Athens had become "modern" in its sense of time. To meet demands for "equal time" in legal pleading and in dramatic contests, it was necessary to have an objectively measured, even-flowing measurement of "time".

At least as early as the fifteenth century B.C. Minoan priests had devised special pierced "libation jars" to keep streams of wine or oil offerings pouring onto the altars. In Greece, by the eighth century, a similar device with small outlet and float was known and used. There is a report that Thales, the inventor and philosopher, had tried to estimate the ratio of the diameter of the sun to that of the entire heavens with such a *klepsydra*. He timed the period it took the sun to rise completely after it first showed above the horizon, then "compared this volume of water to the volume of an entire day".

There were, however, other ways in which time was measured, and other instruments. For an agricultural population what was most important was the cycle of the seasons, with the critical dates of solstices and equinoxes. Calendars, where they gave accurate day and month dates, were inexact and required special sets of "leap days". "Almanacs" measured time by number of days after solstice or equinox, or by the rising of constellations. Anaximander, inventor of cosmic model and map, is reported to have set up a new kind of "sundial" (*gnomon*) at Sparta. But the purpose of his instrument, and that of other early sundials, was to measure the progress of the seasons, rather than the subdivisions of a given day.

However, the shadow of the sundial not only lengthened and shortened with seasonal change, but also could trace out divisions of the sunlight hours of individual days. Taking a "day" as sunrise to sunset, the dial could give a rough division of each day into fractions. Apparently the advantage of this, for civic purposes, was appreciated. In 433 B.C., the astronomer Meton put up a new type of sundial on the Pnyx, the meeting place of the Athenian Assembly. Presumably this official dial defined "meeting" and "adjournment" times.

The *klepsydra*, or small water-clock, was used for more precise measurement. At first these clocks seem to have required frequent refilling. Gradually domestic uses were found for them. As they became larger, a crucial decision had to be made: Was the flow of water, or the daily motion of the sun, to become the official measure of public time? Offhand, one might have bet on the sundial; for this carried with it an overtone of "duplicating the movements of the heavens" and a correlation with actual divisions of a working day that must have made it seem more "natural". But pressure—on court dockets, in overcrowded dramatic contests, in public meetings—was building up demand for objective, fair measurement of relatively short intervals. A six-minute "official" clock, for example, was found in the excavations of the ancient Athenian Agora by archaeologists of the American School.

Apparently, drawing on Professor Derek Price's history of the ancient clock, the deciding factors were the design of new "slow emptying" wells and of a kind of "cosmological model" applied to telling time. A large, smoothly plastered well, with a small metal outlet, could empty slowly but regularly for a period of a day. And if a cable attached to a descending float in the clock-well was wound around an axle at the top, the axle would turn regularly and smoothly. A celestial globe mounted on the axle would rotate automatically past indicator lines marking out the hours.

By the date of Aristotle's *Constitution of Athens*, courtroom time had come under the same public regulation and standardization as other weights and measures. The constitution

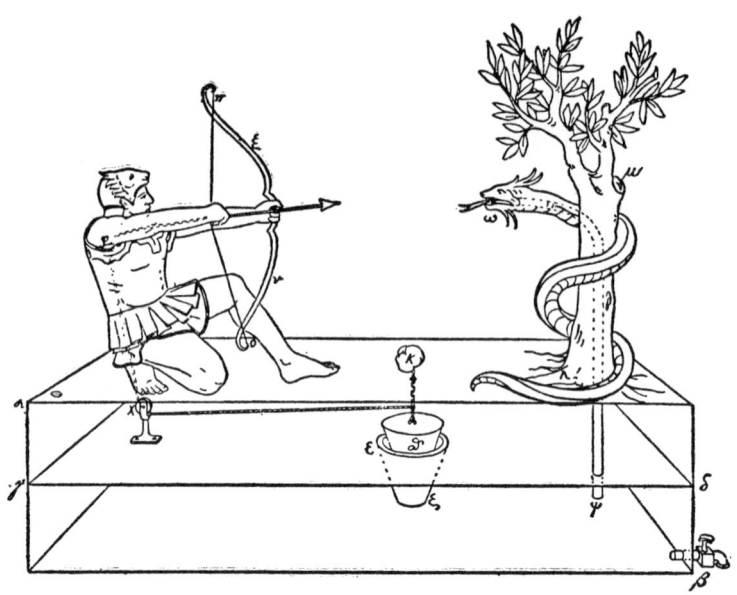

A gadgeteer's toy, designed by Heron. To set it in motion, the apple, attached to the float K, is lifted. This raises the float, triggering off the string attached to Hercules' hand. Hercules releases the arrow and shoots the dragon, which hisses as it is shot.

specified the "number of measures" (of about three quarts each) allowed for pleading in each type of private suit. The number was larger when larger amounts of money or property were involved. The more traditional divisions of "fractions of a day" were still used to define pleading time allowed in cases involving imprisonment, death, exile, confiscation of property, etc. But these were understood as fractions of a "standard day", so that there was no advantage to the defendant tried in midsummer as opposed to winter.

But even before this, the great public water-clock we have described was built along the south border of the Agora. Plato, in his dialogue the *Theaetetus* (written about 360 B.C.)

contrasts the lives of the lawyer and the philosopher. The lawyer lives "driven by the water-clock . . . never at leisure." Already we can see the modern neurosis of clock-watching in one Athenian profession. At the same time, the clock began to dictate to literature. "The length of a tragedy," Aristotle wrote in his *Poetics*, "should not be judged by the water-clock . . . but by what is suitable for the plot." But, under pressure to produce many entries in dramatic contests, the clock became the arbiter. Aristotle's comment is like remarks of some of our modern critics who resent the effect of fixed, expensive blocks of time on the quality of television drama.

After Aristotle, historians generally agree, science as well as politics underwent an important change. As a result of Alexander's campaigns, knowledge of Near Eastern computational astronomy came into the West, modifying the qualitative geometrical models that had been typically Greek. Syntheses of the two traditions led to such achievements as Ptolemy's epicycles, more difficult geometrically and less aesthetic in conception than the older nested spheres, but a more exact basis for prediction and computation.

In the third century B.C., mechanical topics began to be treated in a literary form. Thus a student in the school Aristotle had founded, the Lyceum, compiled a book of mechanical problems (*Mechanica*) which has been preserved along with the works of Aristotle. The book deals with the theory of mechanical advantage, particularly the lever and the wheel, and gives little insight into practical technology or theoretical gadgetry. But it is the first step of mechanical topics toward literary respectability. A few of the problems proposed are relevant to main themes we have been tracing. For example in another notebook the *Problemata*, in the duel between mechanism and dishonesty, the detection of loaded dice is discussed not merely once but twice by this notekeeper. (They revolve in the air when given a "flat" throw.)

And in the *Mechanica* description of direct-drive wheels, which impart circular motion to each other in opposite senses, we find a note: "by mechanism . . . marvels are performed in the temples". That is a reminder of the fact, which we will return to later, that the ideas of "applied mechanics" and "labor-saving devices" occurred to priests long before they did to scientists or to the general public.

In this period philosophy began to settle down into a debate between "schools". As the Roman empire spread and consolidated, the individual seems to have felt lost and uncomfortable in a bureaucratic, impersonal world. With the sense of personal importance and public participation gone, philosophy turned toward doctrines of consolation and escape.

Mathematics was making great advances. Euclid had set a new standard for systematic presentation of elementary geometry; later mathematicians were exploring more remote realms of number-theory, conic sections, and analysis.

Roman technology, with a new notion of practicality, reached a high point of efficiency. But there were also instrument designers and mechanics who continued the older interest of the Greek tradition. Rome and Alexandria, as well as ancient Athens, had their builders of cosmological models, constructors of automata, and makers of toys for adults to admire.

CHAPTER 7

THE AMAZING ARCHIMEDES

In 214 B.C. the Roman general Marcellus, famous for his victories, attacked the city of Syracuse in Sicily. He had a fleet of ships and the latest thing in weapons: a gigantic catapult mounted on the decks of eight ships chained together. The Romans expected to take the city quickly but instead ran into disconcerting surprises designed by Archimedes, whose hobby of applied geometry had included the design of war machines earlier in his career. Before the city finally fell, we are told, the Roman soldiers would run away whenever a rope or pole projected over the city wall, thinking that it might be another of Archimedes' machines.

Archimedes was a strange combination of pure mathematician and inventor. He was one of the few ancient Greeks who had the sort of imagination that could at once see all sorts of practical applications for mechanical principles applied with differences in function and scale.

With an almost modern insight into the application of geometric principles, he designed new types of pump, the screw, presses, block-and-tackle, military engines, and a new self-operating model of the cosmos. His achievements were appreciated, but his genius found no successors. In our study of machinery and gadgetry, he stands alone. He wrote books on arithmetic, plane and solid geometry, but nothing about

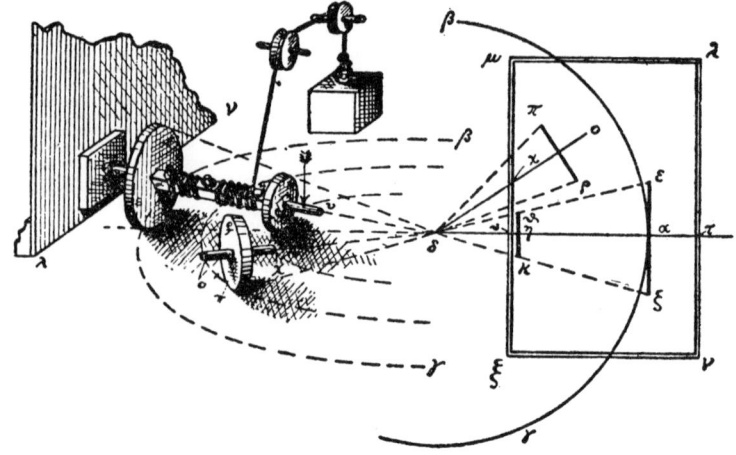

Design for circular motion of an automaton. A weight draws the string from an axle between two wheels. A falling weight draws the string which is wrapped around an axle between two wheels. The automaton runs in a circle because one wheel is smaller than the other. Heron's design includes his demonstration that for wheels producing circular motion the lines connecting the wheels form the cross section of a cone, and that the vertex of the cone is also the center of the circle in which the wheels turn.

mechanics or engineering. He was the inspiration of many stories, making him the first instance of the absentminded scientist turned part-time inventor.

Hiero, the king of Syracuse, was a friend and admirer, and often set his resident genius problems to solve. The most famous of these was the question of whether the king's crown was really solid gold; the difficulty was that the crown could not be melted down to find out. Archimedes was in the public bath, and as he noticed the water overflow when he climbed in, he realized that the loss in weight of an object put into water is equal to the weight of the water it displaces. He ran through the streets of Syracuse shouting, "*Heureka! Heu-*

THE AMAZING ARCHIMEDES 77

reka!" ("I have found it! I have found it!"), with no clothes on.

Having worked on the theory of the lever, he is reputed to have said, "Give me a place to stand, and I will move the earth." That he was not simply joking became clear when he applied the principle of mechanical advantage to a set of pulleys and, singlehanded, hauled a loaded warship along the sand.

It was at this point that the King asked him to design some military machinery; the machines were actually built, and were ready in the arsenal when the Romans came.

Plutarch, the author of parallel lives of famous Greeks and Romans (first century A.D.), gives the story of the siege of Syracuse in his *Life of Marcellus,* quoted here from the Dryden and Clough translation.

> When these could not prevail by treaty, he [Marcellus] proceeded to attack the city [of Syracuse] both by land and sea . . . . Marcellus, with sixty galleys, each with five rows of oars, furnished with all sorts of arms and missiles, and a huge bridge of planks laid upon eight ships chained together, upon which was carried the engine to cast stones and darts, assaulted the walls, relying on the abundance and magnificence of his preparations, and on his own previous glory; all which, however, were, it would seem, but trifles for Archimedes and his machines.
>
> These machines he had designed and contrived, not as matters of any importance, but as mere amusements in geometry; in compliance with King Hiero's desire and request, some little time before, that he should reduce to practice some part of his admirable speculation in science, and by accommodating the theoretic truth to sensation and ordinary use, bring it more within the appreciation of the people in general. Eudoxus and Archytas had been the first originators of this far-famed and highly prized art of mechanics, which they employed as an elegant illustration

of geometrical truths, and as a means of sustaining experimentally, to the satisfaction of the senses, conclusions too intricate for proofs by words and diagrams. As, for example, to solve this problem so often required in constructing geometric figures: Given the two extremes, to find the two mean lines of a proportion. Both these mathematicians used instruments, adapting to their purpose certain curves and sections of lines. But what with Plato's indignation at the method, and his invectives against it as the mere corruption and annihilation of the one good of geometry, which was thus shamefully turning its back on the unembodied objects of pure intelligence to recur to sensation, and to ask help (not to be obtained without base supervisions and depravation) from matter; so it was that mechanics came to be separated from geometry, and, repudiated and neglected by philosophers, took its place as a military art. Archimedes, however, in writing to King Hiero, whose friend and near relation he was, had stated that given the force, any given weight might be moved, and even boasted, we are told, relying on the strength of demonstration, that if there were another earth, by going into it he could remove this. Hiero being struck with amazement at this, and entreating him to make good this problem by actual experiment, and show some great weight moved by a small engine, he fixed accordingly upon a ship of burden out of the king's arsenal, which could not be drawn out of the dock without great labour and many men; and loading her with many passengers and a full freight, sitting himself the while far off, with no great endeavour, but only holding the head of the pulley in his hand, and drawing the cords by degrees, he drew the ship in a straight line, as smoothly and evenly as if she had been in the sea. The king, astonished at this, and convinced of the power of the art, prevailed upon Archimedes to make him engines accommodated to all the purposes, offensive and defensive, of a siege. These the king himself never made use of, because he spent almost all his life in a profound quiet and the highest affluence. But the apparatus was, in most opportune time, ready at hand for the Syracusans, and with it also the engineer himself.

When, therefore, the Romans assaulted the walls in two places at once, fear and consternation stupefied the Syracusans, believing that nothing was able to resist that violence and those forces. But when Archimedes began to ply his engines, he at once shot against the land forces all sorts of missile weapons, and immense masses of stone that came down with incredible noise and violence; against which no man could stand; for they knocked down those upon whom they fell in heaps, breaking all their ranks and files. In the meantime huge poles thrust out over the ships sunk some by the great weights which they let down from on high upon them; others they lifted up into the air by an iron hand or beak like a crane's beak, and, when they had drawn them up by the prow, and set them on end upon the poop, they plunged them to the bottom of the sea; or else the ships, drawn by engines within, and whirled about, were dashed against steep rocks that stood jutting out under the walls, with great destruction of the soldiers that were aboard them. A ship was frequently lifted up to a great height in the air (a dreadful thing to behold), and was rolled to and fro, and kept swinging, until the mariners were all thrown out, when at length it was dashed against the rocks, or was let fall. At the engine that Marcellus brought upon the bridge of ships, which was called *Sambuca*, from some resemblance it had to an instrument of music, while it was as yet approaching the wall, there was discharged a piece of rock of ten talents weight, then a second and a third, which striking upon it with immense force and a noise like thunder, broke all its foundation to pieces, shook out all its fastenings, and completely dislodged it from the bridge. So Marcellus, doubtful what counsel to pursue, drew off his ships to a safer distance, and sounded a retreat to his forces on land. They then took a resolution of coming up under the walls, if it were possible, in the night; thinking that as Archimedes used ropes

Heron's first type of automatic theater performed a sacrifice and dance in honor of Dionysus. As is evident from the figure, the mechanism, shown here in cross section, was concealed above and around as well as beneath the stage.

stretched at length in playing his engines, the soldiers would now be under the shot, and the darts would, for want of sufficient distance to throw them, fly over their heads without effect. But he, it appeared, had long before framed for such occasions engines accommodated to any distance and shorter weapons; and had made numerous small openings in the walls, through which, with engines of a shorter range, unexpected blows were inflicted on the assailants. Thus, when they who thought to deceive the defenders came close up to the walls, instantly a shower of darts and other missile weapons was again cast upon them. And when stones came tumbling down perpendicularly upon their heads, and, as it were, the whole wall shot out arrows at them, they retired. And now, again, as they were going off, arrows and darts of a longer range inflicted a great slaughter among them, and their ships were driven one against another; while they themselves were not able to retaliate in any way. For Archimedes had provided and fixed most of his engines immediately under the wall; whence the Romans, seeing that indefinite mischief overwhelmed them from no visible means, began to think they were fighting with the gods.

Yet Marcellus escaped unhurt, and deriding his own artificers and engineers, "What," said he, "must we give up fighting with this geometrical Briareus, who plays pitch and toss with our ships, and, with the multitude of darts which he showers at a single moment upon us, really outdoes the hundred-handed giants of mythology?" And, doubtless, the rest of the Syracusans were but the body of Archimedes' designs, one soul moving and governing all; for, laying aside all other arms, with this alone they infested the Romans and protected themselves. In fine, when such terror had seized upon the Romans, that, if they did but see a little rope or a piece of wood from the wall, instantly crying out, that there it was again, Archimedes was about to let fly some engine at them, they turned their backs and fled, Marcellus desisted from conflicts and assaults, putting all his hope in a long siege.

THE AMAZING ARCHIMEDES

Thanks to a combination of an overlooked small tower and an enthusiastic festival of Aphrodite, the siege was eventually successful. Marcellus,

> even amidst the congratulations and joy, showed his strong feelings of sympathy and commiseration at seeing all the riches accumulated during a long felicity now dissipated in an hour . . . . But nothing afflicted Marcellus so much as the death of Archimedes, who was then, as fate would have it, intent upon working out some problem by a diagram, and having fixed his mind alike and his eyes upon the subject of his speculation, he never noticed the incursion of the Romans, nor that the city was taken. In this transport of study and contemplation, a soldier, unexpectedly coming up to him, commanded him to follow to Marcellus; which he declining to do before he had worked out his problem to a demonstration, the soldier, enraged, drew his sword and ran him through. Others write that a Roman soldier, running upon him with a drawn sword, offered to kill him; and that Archimedes, looking back, earnestly besought him to hold his hand a little while, that he might not leave what he was then at work upon inconclusive and imperfect; but the soldier, nothing moved by his entreaty, instantly killed him. Others again relate that, as Archimedes was carrying to Marcellus mathematical instruments, dials, spheres, and angles, by which the magnitude of the sun might be measured to the sight, some soldiers seeing him, and thinking that he carried gold in a vessel, slew him. Certain it is that his death was very afflicting to Marcellus . . . .[1]

Archimedes' inventions include the use of the principle of the screw for pumping water to a higher level; the use of burning mirrors as part of his anti-siege machinery; the invention of a self-moving astronomical model, so accurate that it

---

[1] Plutarch, "Life of Marcellus," in *Lives*, trans. Dryden and Clough, Boston, 1857; reprinted in Modern Library edition, New York, n.d. The passage quoted will be found on pp. 376–378 of the reprinted edition.

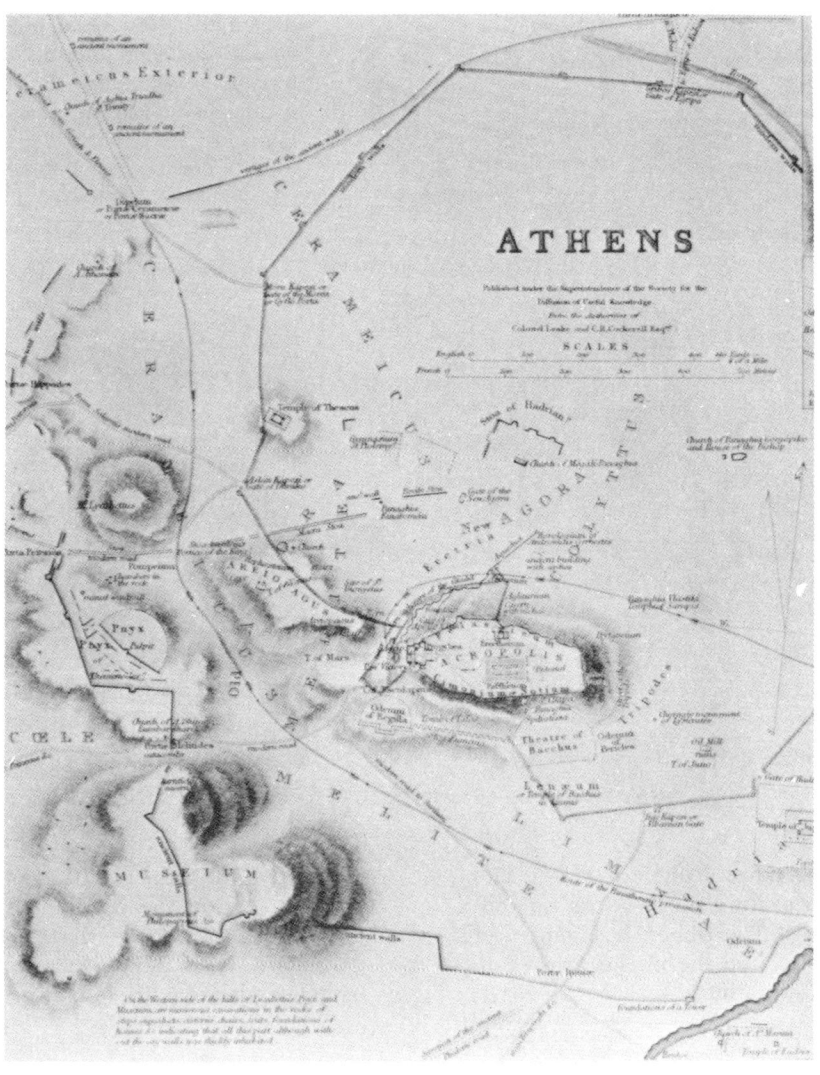

Map of ancient Athens. The Acropolis (*center*) was the city's citadel and center of its religious life. On its south slope was the Theater of Dionysus; just to the north, the market place (Agora) which was the business district. The Pnyx, southwest of the Acropolis, was an adjacent hill where meetings of the Assembly were held. Just off the map to the northwest was the site of Plato's Academy.

Map Collection, Yale University Library

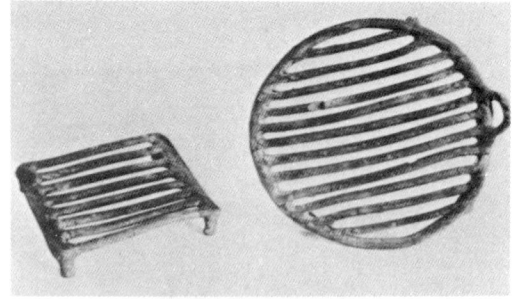

Athenian housewives depended on clay for their utensils. Pottery grills were first built by hand, then fired in a simple kiln.
American School of Classical Studies at Athens

This "novelty pitcher" of the late sixth century B.C. held oil for rubdowns. The ribbon is a symbol of victory, and the pitcher was probably an athletic award.
American School of Classical Studies at Athens

Wine jug, of the late fifth century B.C., found in the excavations of the Agora is in the shape of a woman's head. The rim at the pitcher's top has been broken off.
American School of Classical Studies at Athens

In warm weather wine was served from coolers or *psykters*, such as the one shown at left. The long-stemmed *psykter* was placed in a deep bowl (*right*) which was filled with cold water.
American School of Classical Studies at Athens

"A jet of steam supporting a sphere" is typical of Heron's models, experimenting with steam and hot air pressure. It occurs in his collection just before a design for "the Earth in the Center of the Universe" where, according to Stoic physics, the earth was held at rest by a kind of pneumatic "spiritual pressure."

Studium Generale

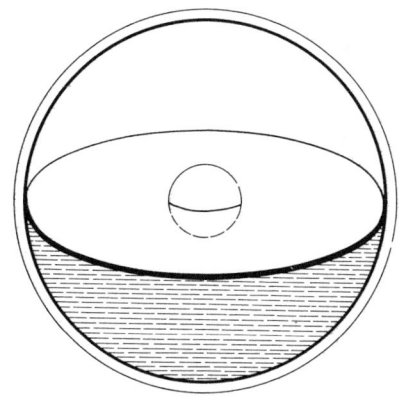

Heron illustrates his concept of the world as the center of the universe with a transparent globe containing air and liquid, in the center of which is a smaller globe representing the earth. A plate of brass divides the two hemispheres, the lower of which is filled with water, the upper with air. As the water is withdrawn, the ball will remain suspended between the two hemispheres. (This drawing, based on the manuscript tradition, has evidently omitted the outlet for withdrawing the water.)

Studium Generale

Modern reconstruction of "the Earth in the Center of the Universe" shows that Heron's model will work when an outlet tube is added. The hole in the plate is made just large enough so that the globe can drop through without sticking. The lower hemisphere is filled with water. As the water is slowly let out, the ball drops into the hole, where a water film seals off the top hemisphere. As the ball sinks farther, the pressure decreases in the upper hemisphere. The combination of pressure difference and surface tension will keep the ball suspended in place. The model (*right*) was designed by Colonel Paul H. Sherrick and the author, and built by E. H. Sargent and Company of Chicago.

Studium Generale

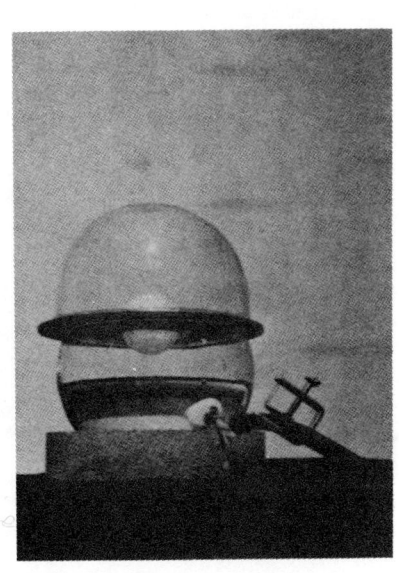

Lead weight of the fourth century B.C. was one of the standard weights and measures that governed Athenian business. The figure stamped in the center is the official seal of the controllers of measures.
American School of Classical Studies at Athens

Three standard measures used in ancient Athens. The official dry measure (*left*) held about 1½ quarts, as did the official nut measure (*center*). The official liquid measure (*right*) held about half a pint.
American School of Classical Studies at Athens

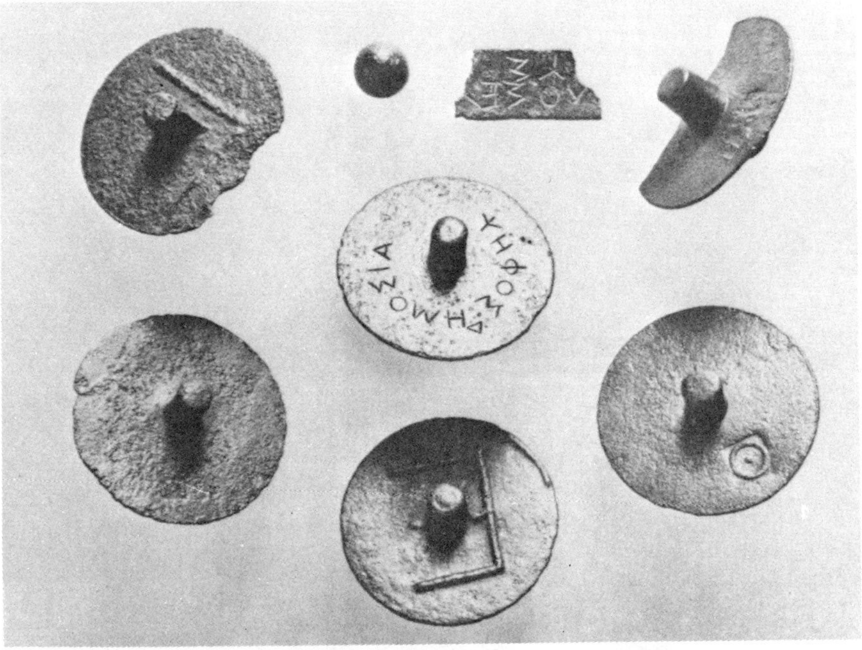

Tickets of wood or bronze (the fragment of a bronze ticket is shown at the top) were inserted into the *kleroterion*, the lottery machine. Then black and white balls were drawn to determine which rows of tickets belonged to jurors who would serve. Votes were recorded by the wheel-shaped ballots, identical except that one had a hollow, the other a solid shaft.

American School of Classical Studies at Athens

Standards were even used for building materials, as evidenced by these standard tiles of the first century B.C. permanently mounted on the market place for the buyer's comparison with his own purchase. Tiles as late as the fifth century were built to the same dimensions.

American School of Classical Studies at Athens

The octagonal "Tower of the Winds" clock still stands in the Roman Agora in Athens. Inside, operated by water, was a large dial that told the hour. Calendar tables may have been mounted around the walls. A weather vane was mounted on the roof. Sundials on the outer walls were added later.

Oskar Seyffert, *Dictionary of Classical Antiquities*

Military machinery designed by Archimedes almost succeeded in keeping the Romans outside the walls of Syracuse. But after a long siege, the Romans took the city. Following the battle, Archimedes was still working on his diagrams when a Roman soldier ran him through with a sword.

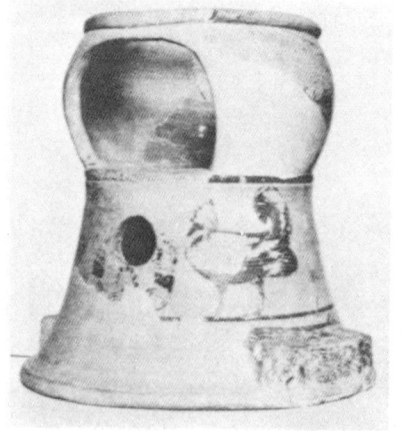

A child's commode, belonging to the sixth century B.C., found in the Agora.
American School of Classical Studies at Athens

The horse on wheels was a popular Athenian toy. This pull-toy, of the fourth century A.D., was made of terra cotta. Its ancestors, less plump but equally mobile, go back 1800 years.
American School of Classical Studies at Athens

Lamps and literature came and went together. Both appeared during the seventh century B.C. and died out fourteen centuries later in the seventh century A.D. Holding a torch in his left hand and a hammer in his right, the god Hephaistos, on this lamp relief, was the patron deity of crafts and of technology.
American School of Classical Studies at Athens

showed eclipses of the sun and moon over relatively short periods (he actually wrote a treatise describing this one); and the engines described above in the quotation from Plutarch.

He had appreciated three things which a modern inventor takes for granted, because by now they are part of our common sense. First, that principles still work with large changes in scale. Thus a small pair of levers that can crack a nut may be built in a size that will hold a battleship; a familiar crowbar, made large enough, could displace the whole earth—if one only had a fulcrum and a place to stand. Second, he realized that mechanical power sources can be transferred from interesting laboratory effects and toys to practical operations. A model of the spiral may, in this way, suggest a mechanism for pumping water; laboratory work with optics and curved mirrors suggested harnessing solar power to a gigantic burning mirror. Third, that there is a kind of logic involved in solving mechanical problems and designing equipment; to duplicate the motions of the heavens by a water-driven model, a long chain of step-by-step combinations of scaled-drive devices has to be visualized and its operation calculated. This is the kind of logic we see at work in the evolution of the lock and key.

Historians today usually tell us that Greek and Roman technology remained undeveloped because their civilizations depended on slave power for useful labor. This meant that the leisure class, and even the middle class, had no incentive to make them think of labor-saving uses of machinery. This explains why the Greeks did not need machinery to save labor, but it does not explain why they did not develop alternative sources of power. The fact is that except for such sporadic exceptions as Archimedes, the inventor's mentality had not yet appeared. By luck and inspired accident, there had indeed been many important inventions in the West. But the mind of the mechanical technician, typified by Heron of Alexandria, whom we will meet in the following chapters, was really that of a gadgeteer.

CHAPTER 8

HERON OF ALEXANDRIA

The two works of Heron of Alexandria, *Pneumatica* and *Automata*, are a treasury of designs marking a golden age of Western gadgetry. Bit by bit, as we have mentioned, the convention that "mechanical" matters were not the sort of thing one wrote books about had broken down. The pseudo-Aristotelian *Mechanica* was the first sign of this change. Archimedes, for all the "disdain for mechanical contrivances" that Plutarch attributed to him, still put into writing a description of his new astronomical model. And, as we have seen, sometime in the third century B.C., an "engineer," Ctesibius, wrote descriptions of an automatic theater and a water-powered pipe organ, and a treatise on military machinery. In the following century, Ctesibius' work was continued—and further written about—by Philo of Byzantium. But we have only scattered fragments of these last two treatises, and none of Archimedes' book. The first important extant source for these achievements of the ancient "mechanics" is Heron.

At one time the similarity between designs of Heron and Ctesibius led scholars to date Heron as early as the second century B.C. But more detailed investigation suggests that he belongs rather to the second century A.D. Heron's references to earlier works show clearly that he sees himself as continuing and perfecting the projects of Ctesibius and Philo when

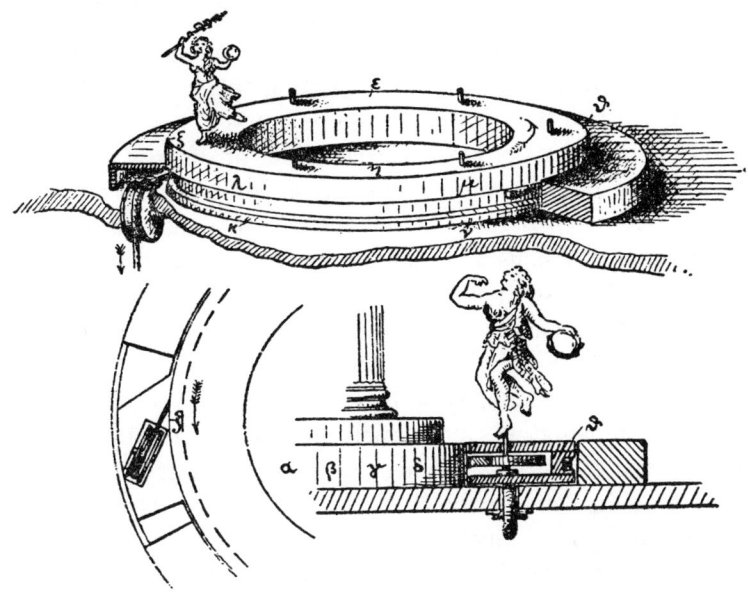

The dancing bacchantes—detail from Heron's first theater. The figures revolved on a turntable on the stage, with each dancer pivoting individually as the mechanism went around.

he does not claim a design or effect as his own invention. We usually assume that he is adopting it from the earlier tradition. We will be particularly interested in the sections of Heron's work that represent culmination of automaton design, a new direction in cosmological models, ideas for using nonanimate power sources, tantalizing relics of an idea of "labor-saving machinery" connected with religion, and one casual contribution to the automation of honesty. These have a double interest. Historically they influenced Arabian, medieval, and Renaissance construction of clocks, automata, fountains, and the like. Retrospectively we see much more clearly through Heron's work what the earlier plans for and attitude toward ingenious mechanism had been. It would be incredible if the more effective tricks of classical designers had not been preserved and used continuously by later mechanics. In fact some of Heron's components probably go back at least to the fifth century B.C.

As Heron's work progressed, the values of surprise and entertainment became more important in his selection of machines. He seemed to have set out to solve the "classical" mechanical problems as much in the spirit of the historian as of the scientist. It is interesting, in this connection, that the theme we have labeled the automation of honesty is relatively unimportant, while toys for adults are much in evidence. The reason may well be simply that Heron, whose professional work was concerned with simplified formulae for surveying and measurement, wanted a holiday. But beyond that, it may also be that the Roman city, as opposed to the Greek small city-state, had already created an objective, impersonal mechanized environment. The mechanically disciplined troops, centralized bureaucratic administration, Empire-wide codes of law, efficient provincial tax collection were, if anything, too objective, as opposed to the Athenian courts and assembly, which had been too personal. In such a world as the Roman, we would expect mechanical ingenuity to move toward amusement, surprise, and escape, rather than toward stiffening the apparatus of administration.

The change from the Greek to the Roman spirit is clearly illustrated in the work *On Architecture*, by the Roman, Vitruvius, in the first century B.C. Vitruvius devoted his tenth book to "machines" and approached the topic with the Roman sense of the "practical." This attitude divides technology into three classes: useful, amazing or comfortable, and irrelevant. "Can it crush gravel and, if not, can we bathe in it?" would be a typical Roman question. Machines for building, for besieging cities, for measuring mileage traveled, and a complex design for a uniform-hour sundial constituted his repertory. Except for a brief postscript mentioning the musical organ of Ctesibius, the automata, orreries, and toys of the Greek tradition are set aside. After reading Plutarch's "Life of Marcellus," it is a particularly effective contrast to see what Vitruvius has to say about defensive military machinery, which he thought was pointless. The best thing to do, he felt, was to flood or mine the ground near the walls, so that rams or towers on wheels would bog down! Roman roads, aqueducts, and baths became part of the common sense of the day. The older Greek theoretical admiration of gadgetry must have seemed out of date.

The design of astronomical equipment seems to have split off from the other mechanical crafts and developed its own technology. In one direction, as we will see, this took the form of a geared computer; in another, water-driven models of celestial motion were built for several centuries, but we learn nothing about them from Heron or from Vitruvius.

The theme of combining cosmology, aesthetics, and utility reappeared in the first century A.D. in the "clock" designed by Andronikos of Rhodes for the Roman Agora in Athens. This is the octagonal Tower of the Winds, which still stands there. The tower was a new and elaborate public clock operated by water. Part of the round reservoir still stands attached to the south wall. The door was kept "open all day and night" so that anyone who wanted to could find out the time. Inside, if Professor Price's imagined restoration is correct, there was a large upright ornamented dial, marked with the hours, on which a hand (attached to a drum, itself fastened to a float) turned. On the roof was mounted a large triton weather vane. Sundials on the outer walls were added later. Calendar tables were probably mounted around the inner walls, and perhaps fountains were fed by surplus water from the clock reservoir. This must have been a useful building, bringing together time and weather information. It combined useful scientific

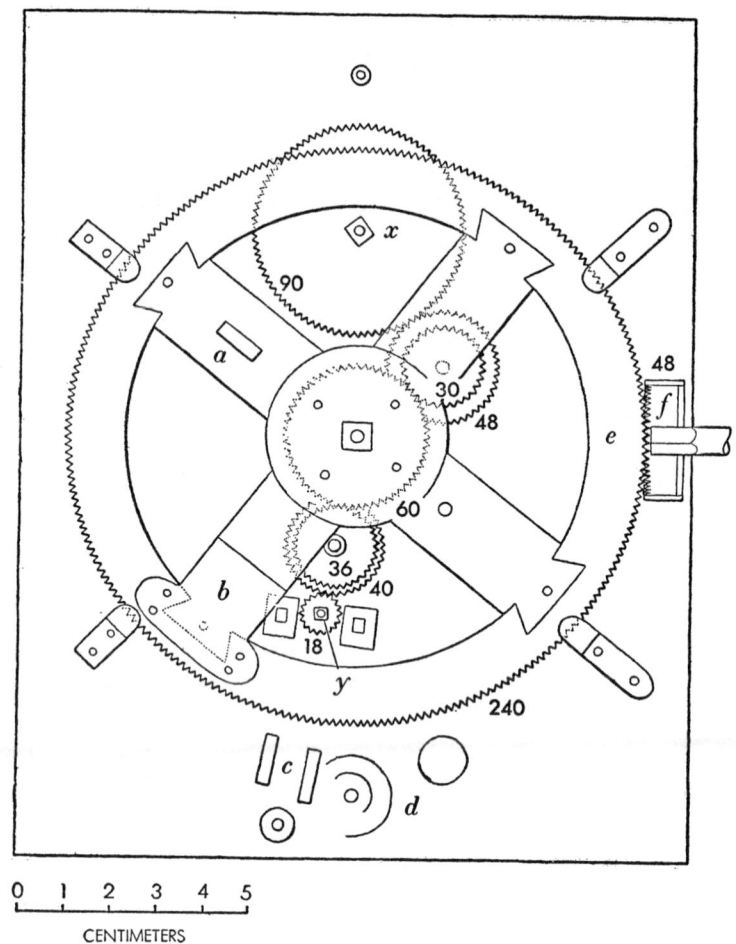

Scientific American and Derek J. DeS. Price

Details of a computer from the first century B.C. Fragments of the computer, found on the wreck of an ancient ship, further supported the idea that the Greeks had developed an advanced technology. It contains a complex set of gears and is believed to have measured astronomical data. The diagram shows a reconstruction of the front view, with numbers corresponding to the number of teeth on each gear.

equipment with aesthetic grace and ornamental effects, and no doubt was admired as a showpiece as much as it was valued for its utility.

But to return to Heron's *Pneumatica:* he begins, as we have noted, with a rather learned scientific defense of the existence of the vacuum. He applied the theory to explain the operation of siphons. Heron then illustrated, with all sorts of mechanical effects using water and air, applications of the theory to machines. This anthology is a classic in the discovery of inanimate sources of power. Ideas brought together here could have inspired a much earlier industrial revolution, given more Roman imagination, less slave labor, and less aristocratic ideas of distribution of commodities. Tucked away in the anthology we also find cases of "labor saving" applications of principles. These appear in the field of "religious technology." The priests, who had to perform certain duties and rituals for themselves, were interested as were their predecessors in saving effort. Their applications of science and technology suggest an extraordinary priestly ingenuity. The outcome of this interaction is the coin-operated slot machine, for which Heron gives a working diagram; but various "miracle" and "divine sign" effects of other types are also recorded.

Heron's write-up does not take account of the special developments in the field of building clocks and moving astronomical models. But without benefit of a literary record, technology in this area had risen to an amazingly high point by 65 A.D. That is the date of a complex astronomical computer, recovered within the past thirty years from a sunken ship off the Greek island of Antikythera. Its restoration and design are described in detail by Professor Derek Price in the *Scientific American* of June 1959. Carefully designed geartrains evidently turned indicator hands on the front dials at speeds which were exact analogues of those of planetary motions. The design and precision are reminders that we cannot judge the achievements of classical technology from literary records alone. Nevertheless, this special branch of machine design seems not to be representative of, nor to have had much impact on, the main stream of invention and technology. That main stream, I think, is represented by the ancient "mechanics" whose accumulated designs appear in Heron's books. As a sign of lack of interaction, in Heron's hundred-odd mechanical contrivances there are only two that make *any* use of toothed gearing. But an exception may have to be made to my general remarks about ancient inventors and machinists for the specialized clock and astronomical analogue computer tradition.

It is not clear how many of Heron's devices are his own invention—indeed, his predecessor, Ctesibius, seems to have built both a power organ and a "magic theater"—but it is clear that he was fascinated by them. From temples, scientific experiments, market, and toys, he brings together the designs that fascinated him: a toy animal that drinks immediately after a knife has passed between its head and body, a pitcher that pours different kinds of wine depending on how heavy a weight is attached to the outlet, an animal that will drink any quantity of water presented to it, and so on. As we mentioned in the first chapter, Heron apparently feels that putting this material into book form is an innovation; his introductory chapter on the vacuum is a gesture toward traditional respectability. From this he can make a natural transition to siphons, as giving experimental proof of his thesis that vacuums exist. Then, from the theory of siphons, he can lead into the devices run by siphon, water, hot air, and steam, where I feel he really is most happy and at home.

The complete contents of the *Pneumatica* show this organization at work, as well as the range of devices Heron described. From this rich store, my own favorites are the steam engine, the turning birds, the earth in space, the coin-operated slot machine, and the section on uses of steam

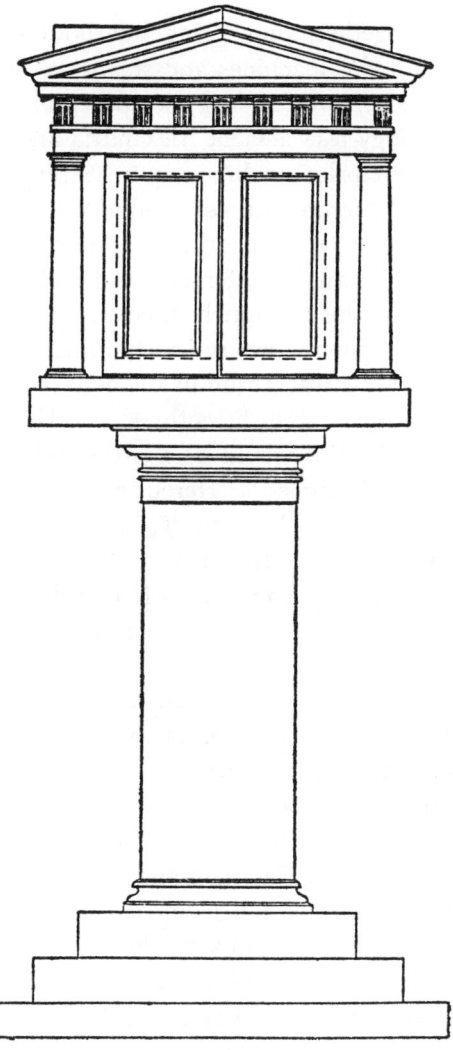

The second type of automated theater designed by Heron had machinery hidden in the pedestal and looked like a grandfather clock. When the trumpet sounded the doors opened, and a five-scene play, *Nauplius,* was performed.

power. Taken together, they bring out both the amazing similarities—with slot machines and steam we are really already in the modern world—and the spectacular differences between technology à la ancient Alexandria and our own.

The coin-operated slot machine requires a bit of context for full appreciation. Heron assumes readers still familiar with the ritual of pagan temples (though not necessarily with Egyptian rituals), and his later copyists and readers have neither this familiarity nor much interest.

In spite of the poor write-up—a combination of offhand Greek and overliteral translation in Bennett Woodcroft's edition—the operation is translucently clear. The coin is introduced in slot *A*—this would be a silver *stater* with a portrait of Alexander costumed as Hercules, weighing a bit over an ounce—and falls on plate *OR*. This, being pivoted on *N*, lifts the valve *S*, so that water flows out through the pipe *LH* and the aperture *M*. The coin slides off, and the weight of the valve again closes it, leaving *OR* back in place for the next coin.

This brings together two components: the lid with a slot was originally designed, as we have seen, to ensure only one vote at a time in the Athenian ballot box; the use of a faucet was an Egyptian temple innovation. The fact that the machine is called a "sacrificial vessel" indicates its function: before sacrifice, in both Egyptian and Greek ritual, the worshiper was required to wash his hands. And so, if we want an inexact but illuminating phrase to describe this early theological application of technology, we can say that the first Western use of coin-operated slot machines was to sell holy water!

In a way, designs for divine signs and temples take us back to the earliest construction of automata. The history of automata in the West begins with moving statues of the gods in ancient Egypt and Crete. The author of the pseudo-Aristotelian *Mechanica* comments in passing on the way miracles are arranged in the temples. Where Heron's pneumatic effects involve temple doors, divine signs, or sacrifices, it seems safe to assume that he is reporting standard mechanisms, though sometimes with his own adaptations. The faucet, the use of steam power, right-angled gearing, and the slot machine attest to an interaction of religion and mechanism that gives a new picture of priestly ingenuity, and a first glimmer of the ideas of applied science for labor-saving machinery.

Heron's machine 31 supplies the intermediate link between the priest running back and forth with a pitcher and the priest at leisure, water-dispensing having been automated. In effect the tank AD is simply our old friend, the Minoan libation jar, but now "concealed behind the entrance pillar" of a temple portico to give a dash of the miraculous to its operation. A close look at the diagram shows that matching holes, bored in the rotating pipe M and the fixed pipe HK in which M revolves will release water when the wheel is turned to an "open" position and will stop the flow as it turns farther. On spinning the wheel, the Egyptian "lustral sprinkling" effect will be produced. But for Greek ritual washing, given a basin, the machine will serve exactly as a modern faucet does.

We can get some further light on this interaction of temples and techniques by looking at machine 17, designed to produce a trumpet sound "on the opening of a temple door". This can be a most effective burglar alarm in case anyone tries to break into the temple to carry off the entire slot machine. But presumably it was really meant as a magical effect. The trumpet is set with its mouthpiece inside a hollow hemisphere, F, resting on the water in the tank $ABCD$. The cord from the door lifts its end of the beam XN (which is pivoted at O) forcing the bell and trumpet down, and thus the effect desired is produced.

I suspect that there were some trade secrets unknown to Heron, among them automatic opening and closing of temple doors. I also suspect that these miracles were originally performed by a concealed assistant moving a lever or pulling a chain that turned the doorpost. In any case, not only Heron, but a tradition of mechanics with a flair for the dramatic, found this automatic door-opening a challenging effect, and set out to build small-scale temples where they could demonstrate it automatically, without concealed assistant-power. Machine 37 is a small temple "such that, on lighting a fire on the altar, the doors shall open spontaneously, and shut again when the fire is extinguished". It is a monster, almost up to Rube Goldberg's mechanical wonders, requiring experts in soldering and plumbing for its execution.

The turning doorposts, with cord and counterweight, probably represent an ancestral miracle design from actual temple use. But in machine 37, the hollow airtight altar is connected by the tube from F to G, with the sphere on three legs below. This sphere, partially filled with water (some mechanics, Heron notes, preferred quicksilver) is, in turn, connected to the bucket X by a siphon, KLM. The fire heats the air inside the altar, forcing it into the sphere. This in turn displaces water from the sphere, which siphons into bucket X; as the bucket lowers, the cords tighten, turning the doorposts. But somewhere in transmission, copyists have forgotten to put a table under X, which is necessary if the doors are to close automatically, for the siphon must operate in reverse when a vacuum is produced in the spherical tank as the air cools. At this point the counterweight at lower right in the design lifts the bucket and turns the doors back closed.

The same effect was to be produced in Heron's machine 38 by a "hot air balloon"—a bladder, inflated by the hot air, rises and opens the doors by pulling on an attached cord. And a further use of hot air appears in Heron's machine 70, "Figures made to dance by Fire on an Altar."

Finally, concluding this line of reflection on technology and temples, Heron's shrine with birds, machine 68, looks much

HERON OF ALEXANDRIA 103

like a sophisticated descendant of the sort of thing the earlier author of the *Mechanica* had in mind when he wrote in his section on direct drive devices, "and this is the mechanism by which they produce miracles in the temples".

When the wheel is turned, the bird revolves and sings. The singing is achieved by the same device as the temple door trumpet, which we have already seen. The revolution of the bird involves use of gears at a right angle, a fairly sophisticated mechanical notion. Again something has happened to the text and probably the diagram in transmission. A modest

A peg-and-lever mechanism produced the hammering effect as a nymph built a ship in the first scene of *Nauplius*. This device matches descriptions of self-moving puppets from the fourth century B.C.

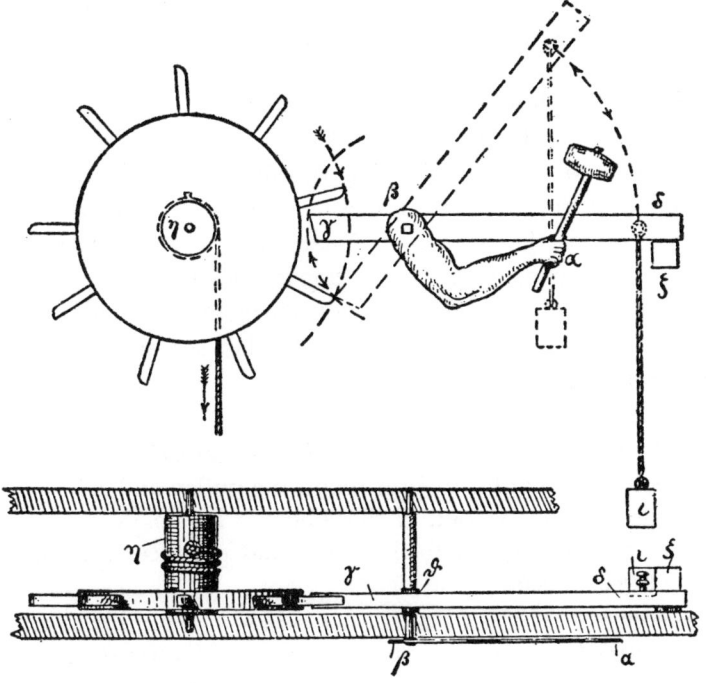

guess concerning the sources that inspired Heron's design may be in order here. He is evidently combining two things: the Egyptian religious use of a bronze wheel, turned as a part of ritual before entering temples, and the Greek use of divine signs indicating that a request or sacrifice is acceptable by a patron deity. His own number 68 is a sort of all-purpose secular model. The blackcap is not sacred to any deity, and the shrine not functioning as omen or oracle.

The textual problem comes in the second sentence of the writeup:

> The construction of a shrine provided with a revolving wheel of bronze, termed a purifier, which worshippers are accustomed to turn round as they enter. Let it be required that, if the wheel is turned, the note of the black-cap shall be produced, and the bird, standing on the top of the shrine, turn round as well; while if the wheel is . . . , the black-cap neither sings nor revolves.

Where the ellipsis points are, all but one of the manuscripts used by Schmidt read "turned," and the young classicist commissioned to translate this for Woodcroft accordingly rendered it "turned [in the opposite direction]. . . ." But the first hand of the Marcianus manuscript read "stationary," corrected above the line to "stopped," finally changed in a later hand to "turned" by some corrector. Schmidt's text reads "stationary." Heron's next to the last sentence confuses the issue further:

> It will be found that, when the wheel HK is made to revolve, the cord is wound round the pulley and raises the conical vessel N; but, if the wheel is let go, N descends by its own weight into the water and produces the sound by the expulsion of the air . . . .

My hunch is that the sound effect was Heron's own addition, rather hastily tacked onto a standard device for bird revolving. What is needed for any divine-sign mechanism of this kind is a *nonturn* setting. And the beauty of the meshed right-angle gearing is precisely that a pull on the wheel or axle, moving the vertical gear, would convert this oracular bird from spin to nonturn.

For several centuries before Heron, a debate had been going on between physical scientists of the Stoic and those of the Epicurean schools of thought. Our records are fragmentary, but we can reconstruct the main points of argument. The Epicureans held that the universe is a concourse of atoms, falling through empty space and colliding by chance. The Stoics held that the universe was made of a continuous matter, differing in density, moved and ordered by a very rarefied material called *pneuma,* "breath" or "spirit," which they identified with God. The Epicureans believed that the truth of theories had to be tested by immediate sense perception—though they were rather common-sense realists than careful experimentalists. Where direct observation was not possible, analogies to available experiences were used. The Stoics held that logical coherence was the essential criterion to be used in testing theories: perception without reason could not be trusted. The Epicureans held that the universe is an infinite stream of particles, falling and tangling; the Stoics, that it is ordered spherical cosmos, with our earth permanently at rest in the center.

S. Sambursky, in his recent book *Stoic Physics,* has gone through the sources and found two passages in which Stoic scientists appealed to experiments against their Epicurean critics. The challenge was that *pneuma,* if it is as rarefied as air (or more so), can't support matter as dense as earth or stone at rest. The Epicurean proof lay simply in an appeal to experience with falling pebbles. Chrysippus the Stoic (third-second centuries B.C.) tried to explain why the earth rests in the center: (1) evidently air, though not dense, can move bodies with more weight; for example, if you put seeds into a bladder and inflate it, they will blow around inside in rapid motion; and (2) a ball, if pulled by attached ropes equally in every direction within a sphere, will remain at rest in the center. These two preserved fragments show us that Stoic physicists did in fact have an interest in experiments defending their view of *pneuma* and the earth at rest against their empirically minded critics; and that the first attempts at defense were not very good—not at all good, for the *pneuma* supporting the earth, unlike the air blowing into the bladder, was by hypothesis at rest; and the strings with equal pull corresponded to no forces recognized in the Stoic theory.

Heron himself was, as we would expect, in sympathy with the Epicurean view, which his chapter on the vacuum supported. But in the background of the different phenomena of flow—water, steam, hot air—studied and harnessed in the *Pneumatica,* he must have felt the tremendous theoretical importance of the *pneuma* controversy. And his strange machine 45, "A Model of the Earth in Space," seemed by its title to have something to do with the matter, though its operation, construction, and even its inclusion in Heron's book at all long remained a mystery.

How did the thing work, and what did it do? In the first place, if "a certain quantity of water is withdrawn", there must originally have been a small outlet from the bottom hemisphere through which to withdraw it. Adding this to the design, and muddling around with a notion that this device, like number 44 preceding it, might run by steam power if a fire were built under it, I managed to enlist the interest of my uncle, Colonel Paul Sherrick. Paul, a professional in invention and equipment design, built a model in his laboratory. After testing it, he wrote me that no tricks of heat or steam were involved in its operation. Just as Heron had said, a pingpong ball floating on the water filling the lower hemisphere would stay suspended by itself in the center hole when the water was withdrawn! And when the model arrived at Yale, this proved to be so.

Precise details of how the model worked, and why, will be found in the article we wrote on this for *Studium Generale*. In general, a combination of surface tension and a slight pressure difference set up as the ball dropped into the hole in the center plate explained the trick.

Very quickly, as news of the world machine in our kitchen, then in my office science laboratory (the bathroom) spread, I saw why Heron had included it. Its behavior delighted and mystified all sorts of people, from engineers to authors of science fiction, and the student paper printed a picture of me peering through the glass sphere with a sour expression.

But where had the creation come from? The machine just before it, the ball supported by a jet of steam, has a family resemblance in its design. And that machine is obviously a more sophisticated extension of Chrysippus' experimental proof that even a rarefied gas can support a solid when the gas is in motion. Suppose, then, that the steam jet had its origin in some Stoic *pneuma* laboratory. On that supposition I can immediately see the function of the earth in space, for here we seem to have in fact exactly what the Epicureans claimed to be an impossibility. For the air, at rest, and theoretically able to escape through the hole (slightly larger than the ball) *supported a solid* in the sphere's center. This does just what Chrysippus' second experiment, which failed, was meant to do—and does it in a way that an Epicurean critic would find particularly annoying. He would, of course, be suspicious of the fact that the setup requires filling with water at the start. But when wet the ball will fall through the hole in the plate without sticking; and obviously the hole will permit air to escape, so that the effect is not a result of compression of the bottom chamber's content by the weight of the ball. Besides, one can show—in fact, must be able to show, if the machine works right—that there is leakage of air into and out of the bottom hemisphere.

The effect is tricky to analyze, and I am morally certain

that this machine and the one preceding it were picked up by Heron from Stoic scientific equipment which fascinated him by its unexpected operation.

There was, as I have mentioned, a great deal more scientific apparatus in ancient Greece than we have ever supposed. But the discovery of this particular effect must have been the result of a happy accident, resulting from observation by someone with the *pneuma* theory in mind. What he observed, I don't know; but some idea of the things that might have given rise to this can be gathered from my article on Plato and the history of science, and Professor Derek Price's discussion of the ancient water-clock.

This is an interesting sidelight on the history of gadgetry if, as I suppose, we here have a case of a scientific apparatus taken over as an entertaining gadget. At any rate, it solves the workings of Heron's mystifying plate, hemispheres, water, and suspended ball.

Heron's book inspired Woodcroft to commission an English translation, and has become famous since the industrial revolution because the fiftieth device it describes, the *aeolipile*, is the first record we have of the steam engine. Heron's adventure with steam is a fine case study of the part that gadgetry and toys can play in the history of invention. Never have the potentialities of a new discovery gone less appreciated. As proof of the point, we have Heron's own section on the uses of steam power, which I will quote below.

The reference in the last sentence of the *aeolipile* description to "the dancing figures" is to the design, described above, for figures made to dance by fire on an altar. This may show more clearly how the ball in the steam engine is driven: the two projecting pipes at top and bottom are bent at right angles, to make escaping steam spin it. In the diagram this does not show clearly.

Later in his account, at number 74, Heron gives a plan for the construction of "A Steam-Boiler from which a hot-Air

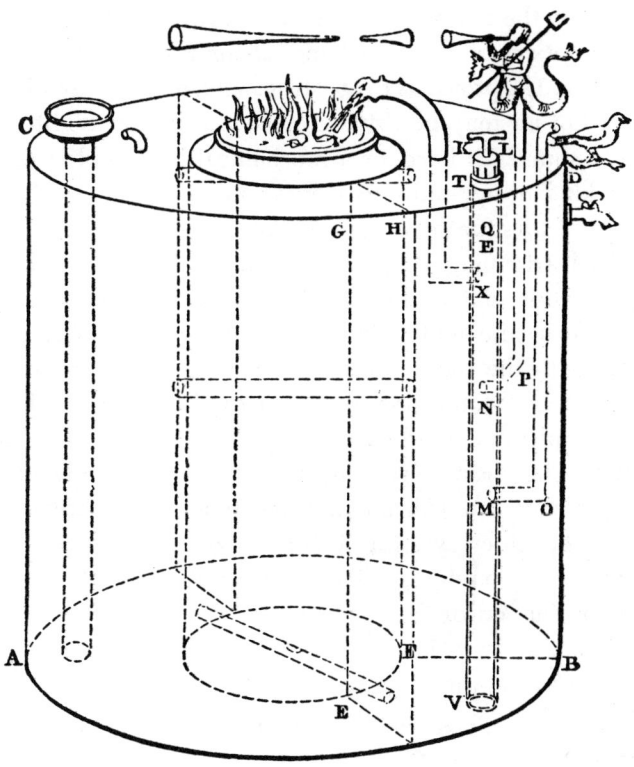

Heron built steam boilers that worked, but used them as gadgets rather than as practical sources of power. The three main uses he could think of for steam power are indicated by the figures on top of his design: a blackbird that sings, a Triton that blows his horn, and a serpent that blasts hot air into the fire.

blast, or hot-Air mixed with Steam is blown into the Fire, and from which hot water flows on the introduction of cold." But what else can be done with a steam boiler besides blow on a fire and heat water? The seventy-fifth write-up is Heron's summary of the applications for steam power that he and his predecessors had found. No better counterweight can be imagined to the temptation a modern reader has to read back our own notions of "inventor" and "steam power" into ancient Alexandria. The project is to construct a steam boiler by which a blast of hot air may be driven to blow a fire, or a Triton be made to sound his horn, or a mechanical bird made to sing.

As a kind of appendix to Heron's miscellaneous gadgetry, some fancy toys for grown-ups complete our picture.

The self-trimming lamp, with its tricky hookup of double rack and wide pinion, belongs here. As the float G sinks, the rack H turns the wide pinion F; and the other side of F, meshed with the rack on the bar DEC, pushes forward the wick which is "wound loosely about the bar." This is, I suppose, completely pointless as a way of keeping the lamp trimmed. The wick, merely loosely wound, won't advance in this right way, but will unwind and tangle; the sliding bar, D, will slip sideways, for it isn't fastened to anything at either E or C. So why is it included here? Evidently because of the fascination of the mechanical hookup of double rack and pinion, an ingenious mechanical idea for which—in the spirit of the "applied geometry" of Archimedes—someone wanted to find an application. As a conversation piece, this probably would have been about as useful as it was pointless as a lamp.

If you own a spring, and want to do something lively with it, Heron's device number 15 may provide the effect you want. This is his design for "Birds made to sing by flowing water; but when an owl turns toward them, they become still". Here Woodcroft's translation and the Teubner text offer two different pictures, the former of which appeared above. As water fills the chamber below the bird bath, air, forced out through pipes, L (disguised as branches on which the birds sit) activates whistling reeds. But when the level reaches the top of the two-tube siphon, FG, the tank drains into bucket Z, air re-enters through funnel H, and the birds are still. An application of the ancient device for automatic temple-door opening makes the effect diverting by synchronizing the songs with the sinister owl. The birds are not deceived by the seeming sweetness in Woodcroft's design or the pensive, incompetent look in Schmidt's Teubner figure. Either a float or a bucket with cable and counterweight will make him turn exactly at the time the bird song starts and stops.

In an appended section 16, needing no diagram, Heron offers a further suggestion for the spring-owner designing outdoor diversion. "In the same manner," he writes, "as that just described the sound of trumpets can be produced!"

But even if you do not own a spring, there are some marvels for indoor entertainment worth consideration. If you like trumpets but lack flowing water, Heron's device 49 offers "A Trumpet, in the Hands of an Automaton, sounded by compressed Air." To work it, you blow hard into the bell of the trumpet. This forces air under pressure into the submerged chamber EFG (a small hole in the bottom lets water escape into the main tank as pressure rises) and, when you are done blowing, the air is forced back out and the bugler sounds his horn.

Or perhaps you prefer "An automaton which will drink any quantity of water presented to it." Heron has a design for this, too. And certainly the liveliest gadget ever devised for cocktail time is Heron's number 32, which he himself thinks of as particularly useful for a "Dutch-treat" wine party. The "Dutch-treat" motif is introduced in Heron's first sentence:

"If several kinds of wine be poured into a vessel by its mouth, any one of them at choice may be drawn out through the same pipe; so that, if several persons have poured in the several wines, each one may receive his own according to the portion poured in by him."

The principle is clear from the diagram. Each weight turns the output pipe a different distance, and holes appropriately bored in this pipe match outlets in the three chambers; a counterweight, K, swings the pipe back to off position when the weight is removed.

Heron's opening sentence describes the operation of his greatest toy, machine 40: "On a pedestal is placed a small tree round which a serpent or dragon is coiled; a figure of Herakles stands near shooting from a bow, and an apple lies upon the pedestal: if anyone raises, with the hand, the apple a little from the pedestal, the Herakles shall discharge his arrow at the serpent, and the serpent hiss."

The pipe Z, with a whistle, does the hissing. The top chamber, ACD, is filled with water; a hole into the bottom chamber which is filled with air is covered by weight H, attached to the apple. As H is lifted, water pouring into the bottom chamber forces air through pipe Z, and the dragon hisses.

But what becomes of the arrow? It is lucky we thought of that. Some misanthrope must have erased an arrow-control unit from Heron's original blueprint, and going unnoticed, this omission might have made the shooting Herakles a more brisk entertainment than we—or Heron—had bargained for.

CHAPTER 9

AUTOMATED TOYS AND THEATERS

As we have shown in chapter three, above, self-moving entities fascinated the ancient Greeks. "The soul is that which can move itself," wrote Plato. But he was only echoing a long tradition. From the magical statues of Daedalus, in the days of King Minos and the labyrinth of Crete, and the early philosophical theories of Thales the Milesian, life and self-motion had been thought of together. Thus the small leather-thong and peg windup toys of the Hellenic Age had been called "the amazing ones" and had fascinated the great philosophers. Mechanism counterfeiting life was the theme that enlisted the talents of Heron. The sense of magic that automata inspired comes through beautifully, in three stages, in his treatise *On Automata*. All the resources of the *Pneumatica*—siphons, pulleys, levers—plus some new ones are combined in this set of the most frivolous and fascinating designs of mechanism duplicating life itself that can be imagined.

Professor Derek Price has suggested that mechanistic philosophy, with its view that living things are complex mechanisms, did not inspire the idea of simulating living things mechanically. On the contrary, he holds, building simulations of natural things in motion was an early project that fascinated human craftsmen, and the more abstract philosophical view came later as an extension. I have not taken sides in this chapter as to which came first, the gadgets or the theory, but the evidence does make clear the fascination that self-moving automata held for the tradition of Greek mechanics culminating with Heron. This fascination, and desire for realistic imitation, are particularly clear in Heron's writeup of the automated theater, where he interrupts his summary of the play to admire the realism of some effects he has designed. Nymphs hammering a ship together and dolphins jumping from a sea are described, followed by an admiring remark: "Just the way they really do!"

The modern reader can supply plenty of contemporary examples. Battery power has replaced the older thongs, weights, and springs of the windup toy, but the design remains the same. My daughter Joanna has, on one toy shelf, the following: A pig who holds a frying-pan with an egg over a glowing stove; he pours on salt, squeals, tosses the egg up and catches it in the pan. A fox with a hat on a stand, wearing a magician's cape; he lifts the hat, and a stream of soap bubbles blow out of it. A family of bears, one of which turns the pages of a book, one irons, one knits, and another drinks in enthusiastic gulps from a bottle. Now in disrepair, at the end of the shelf is a marvelous taxi that steers away from the edge of a table when it is about to run off, known, for reasons not relevant here, as the "Never Go Down Plato Street Taxi." Dolls that sleep, drink, cry, and chat. And there also is an electric train with associated gear in the attic.

The first part of Heron's *Automata* takes up self-moving windup toys. Lacking tension springs, the power source is weights attached to axles. But the effects are a remarkably exhaustive list of patterns of motion. I myself, comparing Heron's work to modern mechanical chipmunks and turtles, am inclined to think that what lent magic to these cord and olive-oil-lubricated dolls on wheels was the mechanical simu-

AUTOMATED TOYS AND THEATERS 115

lation of purpose, life, and freedom. And here is a philosophic point of great importance: Mechanisms that simulate intelligent behavior have, through the ages, sharpened our definitions and appreciation of what is nonmechanical about creativity and life.

Originally, weight-operated dolls and theaters had one fatal mechanical limitation: the weights fell too fast. Thus, I am told by my colleagues in classics, the first automated theaters, where actors were run by weights in a state of free fall, required a cliff for their production. These early cliff-hangers, even at Delphi or Rome, were perforce brief, and mechanically crude as well, since the force of the falling weights tended to jerk any but the most wooden actors offstage and over the backstage backdrop.

By Heron's time, an application of the hourglass had set that right. The weight could be mounted in the top of an hourglass-shaped wooden frame, filled with barley seed, which ran slowly through a small aperture into a lower bin. As the level lowered, so did the weight, slowly reeling in the twisted line—of flax, wrote Heron, since bow-strings stretch and spoil the show. (The gravity-drive weights can be seen in the designs for the automatic theaters; an exactly similar "motor" was used for the automata with wheels.)

It now remained for the self-moving "wonderful toys" to duplicate the patterns of living beings following their own impulses. Heron, with the inspired objectivity of the mechanic, and the subjective tonality of Greek humanism, thought that there were three relevant types of "wonderful mechanical motion." These were the sorts of thing that individual entities—race horses, planets, small beetles—could be found doing in nature. Whether he felt he was resolving the mind-body problem progressively for persons, stars, and small darting scarabs and mayflies is not indicated in his writings. But with an inherited classification of kinds of motion, he does set out to duplicate each one with his small wheeled contrivances. A great deal of talk had gone on about the beauty and regularity of circular motion, either simply repeating or curving on into a figure 8. Ethics and mechanics had long recognized that the motion of maximum efficiency —whether for realizing a desire or obeying a technological law—was to connect two points by a straight Euclidean line. The Epicurean version of the atomic theory had focused a spotlight on motion of quite another type—an unpredictable, right-angled *kline* by which atoms falling in parallel lines suddenly and unpredictably "swerved" like darting insects, collided, tangled, and created a world out of what would otherwise have been a mere infinite chute of particles in parallel routes of fall. I suppose that this sort of right-angled dash has always been the most "amazing": rats, colliding fellow pedestrians, perverse bouncing balls, all have it. And the mechanical logic by which Heron simulates the swerve is a feat of ingenuity that shows how far the technique of thinking with mechanical logic had come from the Homeric door fastening by latchstring and hook.

Heron's gravity-drive for wheeled devices is shown in the figure. The weight at the top is attached to the axle by a flax cord. The axle and wheels are a single rigid piece, turning in the pivots *epsilon* and *xi* at either end. In the box is placed sand—or better, since it is lighter in weight, whole grain— which pours gradually into the lower chamber when the small aperture is opened. As the weight falls, the tightening cord turns the wheels in the direction indicated by the arrow in the diagram.

For the first set of designs "for automata made to move forward in a straight line," a third wheel is mounted in front of the driving axle as shown in the figure. Heron, or his later annotators, are leaving as little as possible to chance. Like any good do-it-yourself manual, the *Automata* has diagrams all the way for each "fit flap *A* into slot *B*" step.

"And to make the automaton reverse its direction backward is simply done, as will be shown." Again come the figures. The top and bottom are two ways of winding the cord around the axle for simple reversal at a prearranged distance. The middle scheme takes advantage of the slow fall of the weight to program a stop of any desired time before the reverse motion commences. A further accompanying drawing of motor and chassis is added to show how the machine is driven in reverse.

Heron now moves on to his second problem. "Motion in a circle can be designed in the manner following." The trick this time is to make the wheels of unequal sizes, such that they will trace concentric circles. Again we are given the diagram, this time with a mathematical demonstration that the wheels must equal cross sections of a cone with its vertex at the center of the circular path. Circles of any desired (practicable) size can be programed by proper wheel adjustment. This is too elementary to satisfy Heron, who will move presently from the three "elementary" motions—linear, circular, and rectangular—to "composite" ones, completed automatically without clumsy pauses while the mechanic takes the machine apart to put other-sized wheels on.

Although "travel in a rectangular path" counts geometrically as one of the simplest motions, we have already noted that in nature it is the most surprising, and it is the most mechanically complex. Heron's problem has obviously occurred to every motorist who has tried to park a car in a space of exactly its own length. The only driver I ever knew who solved the problem, getting his car sideways in a garage where it touched both sides, had to take his garage down the next day because he couldn't remember how he had done this the night before. I suspect that if Heron were alive today with a patent on his solution, he could sell rights for a fortune in Detroit.

However, what I most admire is not the mere utilitarian application but rather the mental agility used in designing

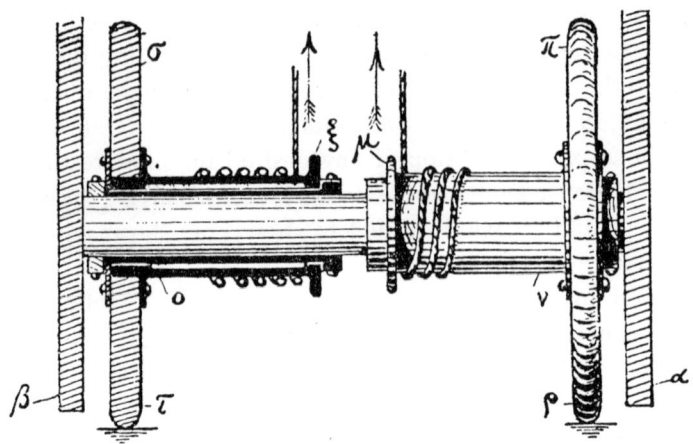

Axle design for complex motion of automata, where each wheel turned independently of the axle and of the other wheel, anticipated the modern automobile differential.

mechanisms to perform a given unlikely operation. The first step is a recognition that with weight and fixed axle drive, one pair of wheels cannot possibly do the trick. But two pairs at right-angles aren't going to work either; the axles can't cross and one pair will always drag. Yet, reasoning from the wheel designs used to produce circular motion, it is clear that this right-angled wheel mounting could be done *if* the wheels differed in size. Thus the figure shows two pairs of drive wheels and two front wheels of different sizes mounted at right angles in the chassis. But how are the wheels to be changed? Here an earlier Mycenaean or even Athenian designer of toy chariots would have been baffled; equipping each toy with a slave to lift it and change axles would have been the only theoretical solution they could have found, and this, at least in Athens, would have priced the toys out of the market and spoiled their enchantment. Obviously, to change drive, the large wheels must be down, lifting the smaller ones

AUTOMATED TOYS AND THEATERS 119

out of the way. Here, as the figures show, Heron takes advantage of Archimedes' work on the screw and introduces worm gearing, for the first time in a Western machine as far as I know. The fact that the wheels must press down hard enough to lift the whole machine, and that a screw-press was invented about 75 B.C., suggests the association of ideas that led to this adaptation. A close look at the diagrams shows that the large wheels are mounted in moving parts, which raise and lower as the falling weight transfers its force to the string turning the worm gears; the support for the moving wheel-mounting rests in the grooves of the gearing and thus the whole assembly will be forced down. (See page 54.)

At this point, some general doubts concerning the practicality of Heron's designs occur. I am sure—for reasons mentioned above—that he had actually built, or at least observed in operation, machine 45 of the *Pneumatica*. His description of the design of the six-wheeled chassis for rectangular motion has the same mechanical verisimilitude—what sorts of oil, cord, weight to use, what parts must not be too mechanically tight—as his descriptions of the much simpler devices that he could have built. But the more I look at the tongue—worm gear—sliding pivot mounting in this figure, the more I think that this is a purely theoretical solution to the mechanical problem that intrigued him. Detroit will have to go back to the drawing board before the new stem-winder models will actually park by rectangular motion, I'm afraid. In the first place, as soon as the large wheels touch the ground and begin to lift the whole weight of the machine, that single metal tongue which is pressed down will force the sliding panel sideways, and jam it. In the second place, one has some doubts as to how the complex detail of cord-drive can be fitted together in the finite interior of the moving automaton. True, unlike Detroit, the conditions of Heron's problem do not require room to be left clear for passengers or cargo; but notice that he needs separate driving cords for the first axle, then for the flywheel synchronizing the screws, from this to the screws and, finally, for the axle carrying the large wheels once they are lowered into position. The odds seem to me high that the actual interior maze of whizzing strings and spinning wheels would lead to a topologically interesting, but mechanically nonfunctional tangle.

Part one—self-moving windup toys—concludes with an equally ingenious and much more practicable device for presetting various types of complex automaton motion. As the usual schematic figures show, this time the two wheels are not fastened to the axle, but turn independently, each on its own shaft. Should we credit Heron with the invention of the differential here? In any case, the separate windings of the two half-axles will evidently make possible new and fantastic feats of behavior: S-curves, turning on an obolus, figure 8's ("horse-fetter curves" in Heron's vocabulary). In principle art can now imitate any or all the motions—purposive or random—found in nature.

A final redesign of the chassis, introduced to make the wheels "drive the automaton much faster and freer" concludes this phase of automation.

"Concerning the motions of this type of automaton, we have said enough", writes Heron, and moves on abruptly to "this is the way we will cause a fire to kindle on an altar." The transition seems surprising, until the next elegant figure shows us that the altar in question is part of the equipment being combined for a performance in an automated theater. This is the theater "of the first type" in which three-dimensional moving figures stage the performance. Heron indicates as he goes along that there can be a number of variations, but the outline of what happens is given by the following scenario:

Dionysus stands, cup in hand, before an altar. At his feet is a tiger: a Nike surmounts the roof of the temple. Around the temple stands a circle of Bacchantes. [A maze of siphons, pipes, cords, and pulleys are housed in the hollow columns and in hidden top and bottom compartments of this stage set.]

There is a sound of cymbals and trumpets. At once, as Dionysus turns toward the altar, a fire kindles. From his cup a libation of wine flows. The Nike on the temple roof revolves. [The tiger at some point here drinks milk, with variations suggested.] The god turns again: there is a peal of thunder, and suddenly the Bacchantes dance around the temple in a circle, each dancer pivoting and pirouetting constantly as they go around. Finally, the performance stops, and the tableau is set exactly as at the beginning.

Particularly for an audience familiar with the tableaux which were part of the revelation made to the initiate in the Mystery religions, this is an attractive program. As a matter of fact, apart from improved sound effects, we would admire it today—if there still were lovers of the automated theater, with its wooden actors. Maybe we have grown tired of wooden actors through seeing too many. A colleague of mine cruelly suggested that Heron's play for wooden silhouette figures was at about the right level for the talents of a local amateur theatrical group he had just been watching. More likely, our sense of romance has followed the industrial revolution. Now model railroads, rather than model theaters, hold a magical fascination, as one tries to get whistle sounds, mail pickup, automatic switching, gaudy high-speed motion, and real smoke from the model engines.

But there are still people interested in imitating living creatures mechanically, and in applying modern technology to automated theater design. A recent Sunday television program showed Walt Disney explaining and demonstrating his "Tiki Room" at Disneyland. In this room, mechanical birds fly, sing, talk, and breathe in lifelike synchronized ballets and conversations. The mechanical key is an elaborately programmed modern computer. This is a splendid application of our new computer science, and one that would not occur to most technicians in the field. It reminds me a bit of chapters on uses of steam power and realistic jumping dolphins in Heron. Just as Rube Goldberg's inventions recapture the spirit of mechanical logic at an early experimental stage, Walt Disney's fantasy recaptures the spirit of the designers of automata who were so fascinated by the simulation of animals, actresses, and storms at sea.

Heron proceeds with the same enthusiasm I once had for an automatic milk-can-loading toy train car, to get the exact realistic effects his program envisages. The principles are all familiar already from the *Pneumatica;* what is new is their programmed combination, though there are one or two technical innovations that Heron specially commends to his reader's attention.

In the cross-section figure, we can spot at once the milk and wine storage tanks, the weight-and-cord drive that spins the Nike and pivots the statue of the god. The automatic on-and-off faucets below the stage at the right are more ingenious; if they really worked, they amounted to timed on-off devices, and solved the problem of a revolving platform connected, when wanted, by rigid pipes to a fixed reservoir. It is a nice idea, though a revolving tub or showerbath, which is the only practical application I can think of at the moment, still would be too exotic for the current plumbing market.

Two of the special effects Heron particularly liked were the sudden kindling of the altar fire and the sound of thunder. This depends on a hollow altar, with smoldering fuel inside, and a top plate that is suddenly pulled aside by an attached chain or cord. When the plate is removed, smoke and fire will pour out from the hollow center. The sound of thunder, though, had never been satisfactorily worked out. Philo of Byzantium, whose program for the temple performance Heron is following, called for this effect in his script, but didn't indicate the mechanism he proposed to use to produce

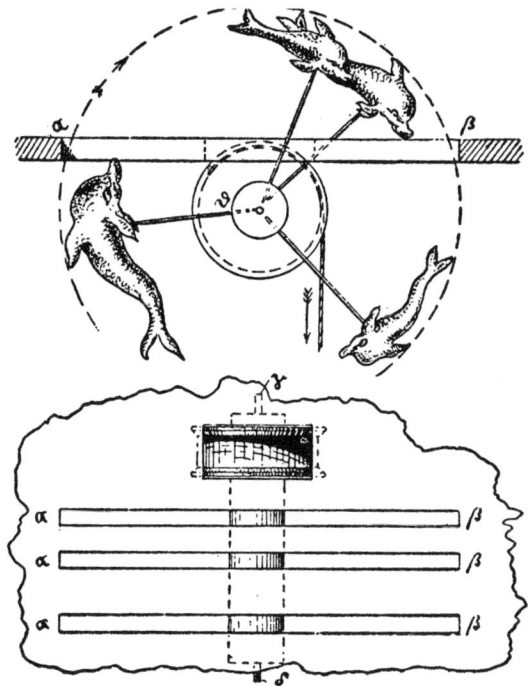

Leaping dolphins, one of Heron's favorite theatrical effects, appeared in scene three of the play *Nauplius*. The dolphins, turning on a pivot below stage level, seemed to jump from the ocean as they swung above stage.

it. The Teubner editor could not resist a footnote to the figure: "This Figure contains essentially the way we make thunder sound-effects in our own time." Essentially, except for the baffles and the tin trough at the bottom, what we have here is the balls, tube, release, and hopper of the Athenian lottery machine.

Finally, we come to the pivoting and circling Bacchic dancers. The way they work is so clear from the figure, and so clumsy to describe in words, that no comment other than a commendation seems called for.

We turn now to a second type of automated theater. In this

one, the stage doors swing open and close again to mark successive acts. The spectacle is two-dimensional viewed against a backdrop at the back of the stage. Nymphs, a ship, Ajax, and others run on tracks, controlling their entrances and exits. The play, traditional for mechanics building magic theaters, is the *Nauplius*. The technique, notice, is the closest thing to our modern motion pictures that the West had until invention of the magic lantern, attributed to P. Kircher in the seventeenth century. And the marvel is that the whole performance takes place automatically: music, scene changes, disappearing ship, and all.

The theme of the play is a part of the story of the Trojan War. The son of Nauplius, king of Nauplia, was Palamedes, who accompanied the Greeks against Troy. A quarrel having sprung up, Odysseus and Ajax the Locrian ("Little Ajax," as opposed to the giant Ajax, son of Telamon) arranged a false accusation of treason which led to the stoning to death of Palamedes. Nauplius called upon Athena for vengeance; and, as the ship of Ajax was passing south along the east coast of Mykonos, the godess sent a storm which destroyed it, and in which Ajax drowned at sea. His body was recovered and buried on the shore by the crew of Odysseus, according to one version of the legend.

Here is the play.

NAUPLIUS: A PLAY IN FIVE SCENES,
FOR THE AUTOMATED THEATER

At the beginning, the curtains open, then in the picture there appear twelve figures, arranged in three rows. These represent the Danaïds, who are repairing the ship and moving it forward to be launched into the sea. These figures move themselves busily: one is sawing, others are hammering, while yet others work with large and small hole-boring tools. And there is a great noise, as of the sound of actual working.

After a predetermined time, however, the doors close, then opening again another scene appears. Namely, one sees how the ship of the Achaeans is launched in the sea.

After the doors have closed and opened once more, one sees nothing on the background except empty sky and painted sea. After a short time, the ships sail in a line; as one disappears, another comes into view. Often dolphins swim along with them, which quickly dive into the sea, then become visible, just as they really do. Shortly thereafter, the sea becomes stormy and the ships run with sail close-hauled.

When the doors have once more closed and opened, nothing of the ships is to be seen, but there stand Nauplius with a raised torch and Athena, and fire burns above the stage, as if the torch were casting its light above.

Closing and opening again, there appears the shipwreck of Ajax's boat, and Ajax swimming. A machine raises Athena above the stage out of view; thunder crashes, and a bolt of lightning falls directly from above the stage onto Ajax, who is made to disappear. And in this way the story comes to its conclusion, as the stage doors close.

This is the end of the five-act play, and of the legend. Heron is nervously impatient to get on with the machinery that makes the performance possible. We can recognize his hand in the incorporation of several effects already utilized in the theater with moving three-dimensional figures. But I would like to pause just for a moment to remark that if you ever are on Mykonos and go away from the tourist center to the eastern coast of the island, you will see the ruined gateway of a Frankish fort. At its base lie great polygonal stones, some thriftily rearranged for a sheep pen. And here, at the point the islanders call the *Taphos Iantos,* they say the crew of Odysseus buried the body of Ajax the Locrian. It is a long way from the gentle parlor magic of an automatic theater!

But it is a challenge to the theatrical realist—for spectacle, the rocks and storms of the Aegean are unbeatable. My sympathy with Heron grows when I note how often *kathaper tes alethes,* "just the way they really are", breaks into his scenario, from the busy shipbuilding girls in scene one, through the diving dolphins, to the torch of Nauplius.

A long section, with diagrams of weights and complex axle-windings, describes the timing for opening and closing the doors, and lowering successive new backdrops for each scene. There follow directions for the Danaids with hammers, who do such a realistic job of ship-fitting in scene one. The attached diagram makes the ingenious hammering motion clear. It also seems to me to make clear one of the designs for automata, in which "pegs striking rods make them move," that so impressed Plato and Aristotle.

The next figure that catches my fancy is the design for the dolphins, leaping from the sea "just as they really do" for the third scene. The three fish, mounted on a turning wheel below stage level, are a very nice effect. But, like the pegs and rod, they are not an innovation, but the application of a very ancient toy design. Coin enthusiasts may remember the *triskeli*—three legs with bare feet mounted on a central pivot—that figure on the back of an Aspendos tetradrachma of the fourth century B.C., and even earlier, I am told, on some Sicilian coinage. If you will visualize these three spinning barefoot legs, you can see why I would bet my last dollar that, before this design caught the fancy of die-engravers of Greek coins, it had already caught the more practical fancy of some toy manufacturer. If you want a barefoot push-toy athlete whose feet really fly as he runs the one-stade dash, this is the best way to build him; if you want a coin design, this one would never occur without the suggestion of the push-toy.

The fiery glow above the torch of Nauplius is Heron's favorite device for having a fire flare up from an altar, as a metal plate above smoldering coals is pulled aside by a control cord.

Finally, the illustrators of Heron's manuscripts—bypassing designs I would have enjoyed seeing for the swimming Ajax and the airlift of Athena—take up the greatest effect of all: the bolt of lightning, an instantaneous flash, after which the figure of Ajax has completely disappeared! With all due respect to tradition, this and the thunder crash are, Heron insists, his own improvements on the unexplained and probably unsatisfactory lightning and thunder in the original automated *Nauplius* designed by his predecessor, Philo of Byzantium.

This time the diagram is *not* self-explanatory. We learn from the text that the lightning flashes first down, then up again, while its motion hides the unrolling of a new backdrop on which the spot occupied before by Ajax is replaced by a patch the uniform color of the sea. The relative slowness of unrolling the backdrop explains why Heron was so pleased with his flash of lightning lengthened by having it go up after shooting down, I suppose. But owners and hired copyists of Heron's manuscripts became more concerned with windup toys—and, presently, with marvelous striking clocks—than with the one-play theater performing without actors or manager. The final version, reproduced here, was drawn—as our other illustrations, reproduced from the Teubner text also were—by the copyist of a fourteenth-century manuscript, now in the great Marciana Library of Venice. So far, apart from butchering the "nonturn" feature of the revolving birds in the *Pneumatica*, he had done rather well; and one can excuse a Byzantine monk for failure to grasp nuances of pagan theology.

But this time, the figure will not do at all. The fact that it has no Greek letters indicating moving parts suggests that there had been some erosion between Heron's original schematic plan—prizing this effect as he did, we can be pretty

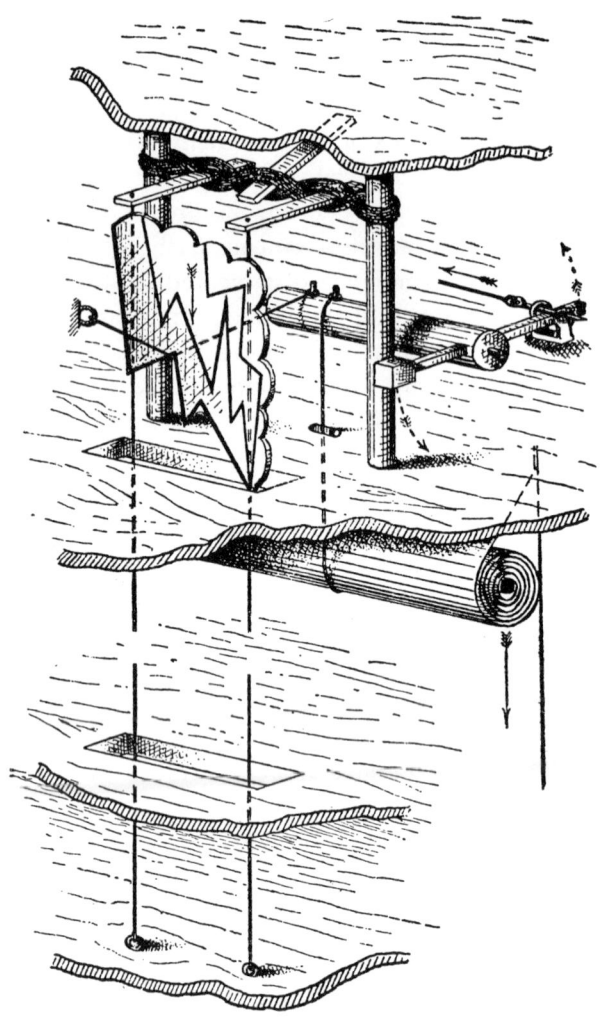

Heron's favorite stage effect: the bolt of lightning. By using twisted cords, the bolt is dropped and then pulled up again. This makes the flash long enough to hide the changing to a new backdrop, and getting the figure of Ajax swimming out of sight. Heron's copyists did not share his enthusiasm: the original lettering of parts has been lost, and the mechanism miscopied so that the present design would make the lightning flash a complete failure.

sure it was the handsomest blueprint of them all—and this reconstitution. As drawn, the operation would begin in the upper right-hand corner, where the pin retaining the weighted arm is pulled out to the left, as shown by the arrow. Fine. Next the weight drops and the arm raises, as two arrows indicate, and the axle through which the arm (with the weight at the front) passes turns toward the front as the weight falls. At this point, the cord passing from the right-hand peg on the axle, going through a slot in the stage ceiling to the rolled backdrop which is painted to replace Ajax with a spot the exact color of the sea, loosens. As it does this, nothing happens, for the backdrop is also fastened by another anchor cord (far right, partly a dotted line) to a fixed vertical pin in the ceiling, making the arrow pointing down just below the roll at the right indicate a mechanical impossibility. Perhaps it is the cord running from the left-hand peg on the axle, beneath the lightning to a retaining pin in the side wall, that does the trick? Not at all. The forward turn of the axle also slackens this lightning control. The pin stays, the lightning sags, tips of it dipping into view onstage—and the intended finale is a fiasco.

But, to rescue Heron from illiterate later admirers, we need only to replace one slot by a pin, and rotate the axle the other way. As it now turns *backward* (the weight is now behind the main shaft), let two cords fastened to the right-hand peg pull out two retaining pins that hold the backdrop. At the same time, the cord from the left-hand peg pulls the "lightning-release" staple at the far left in the diagram. The lightning falls, winding the twisted cord tighter, and then, having passed in front of Ajax, is snapped back up again. The new backdrop has unrolled into place, and the swimming figure disappeared.

And with this rescue of Heron's prize effect from inept manuscript copyists, we conclude the discussion of the automatic theater.

CHAPTER 10

THE INVENTIVE INSTINCT

THE TOOLS for an industrial revolution and a mechanization of civilized life were already on hand by A.D. 100. The sense of mechanical logic, largely lacking in Mycenae, is now fully developed, as we can see from Heron's elaborate chains of action in his more complicated toy designs. The modern lock and key, with fixed wards to prevent the wrong key from turning, had also evolved by this Roman time.

All that is lacking, now, is the idea of using power sources that work for toys on a large scale to perform useful functions. Steam, wind, hot air, water, and gravity power all are available to the ambitious engineer. With Heron, we seem already close to our modern age, complete with steam engine and slot machine.

The automation of the puppet theater, a pointless project undertaken with great enthusiasm, had been complete in principle when Heron discovered a successful way to produce the lightning effects that his predecessor, Ctesibius had fudged in *his* design.

The ancient idea of building self-moving models of the heavens also came to fruition in the second century A.D. A mathematics book of the period mentions the harnessing of water power to driving models of this sort. The idea of specially designed scientific demonstration apparatus had emerged. We can see two examples of this in the Stoic apparatus preserved as Heron's "The Earth in Space" and in Heron's own use of simple siphons to demonstrate his thesis of the existence of a vacuum.

Even more a sign of the change in the world from ancient to modern was the Andronikos clock (the Tower of the Winds) in Athens. Gadgetry still goes on, with ornamental pins, novelty pitchers, grotesque lamps, miniature ceramic animals, and the like, including the entertainment of watching a small ball spun by steam.

From the stone age on, there has been some impulse in human nature to experiment with new styles of tools and artifacts. The record shows miniaturization, new combination of design and function, multipurpose single implements, calculation of more extensive mechanical sequences, and apparatus designed to duplicate nature, or put questions to her.

It had been generally accepted until the present century that both technology and gadgetry were late arrivals on the Western scene. The old picture of Greeks and Romans gave us leisure-class gentry, incurious about and contemptuous of arts and crafts or mechanisms of any kind. Twentieth-century archaeology and work in history of science have shown that this was a mistaken picture. The reason for its development was a double one: the neoclassical picture of the ancients denied them color and flamboyance altogether, in the interests of a postulated austere purity; the ancient writers themselves did not think of technology as the sort of thing one wrote about, so that there is very little literature. The truth is, as I have tried to indicate by a set of samples, that invention and gadget-design gradually blossomed to a nearly contemporary fruition by the second century A.D.

As we have seen, there are four particularly important points of interaction between this rediscovered technology and the history of Greek science and philosophy. The first is the development of the atomic theory by Leucippus and Democritus in the fourth century B.C. According to that theory, all reality consisted of uncuttable (*atoma*) particles of matter, moving in empty space. All change was explained as reducible to transfer of momentum, with the corollary that men and animals were themselves merely complex mechanisms within the larger mechanism of a cosmic world machine. This view has plausibility only when there already are enough ingenious "mechanisms" known to a culture to make the suggested extension of mechanical principles to beings with complex behavior intuitively plausible. One is much more likely to accept this theory after having wondered at the self-moving marionettes of ancient Athens or the hissing toy serpents of Alexandria. And from its inception, I am sure the atomic theory depended in this way on the tradition of invention and gadgetry that ran parallel to it.

One case of seemingly direct action of theory on machine design is the ballot box of the Athenians. The atomic theory had explained why, for example, light but not water would pass through glass, by postulating small openings, "pores," running through the glass. The fire particles were small enough to pass through, but the size and shape of water particles prevented them from so passing. This idea of matching particles and pores goes back to at least 440 B.C., and the ballot box with its exact ballot-shaped slot is almost certainly an applied version of this older idea.

A second point of interaction was mentioned above in connection with the many attempts at standardization of commerce that left their traces in the Athenian Agora. The Sophists, traveling teachers of culture, advanced the thesis that social standards are *not* expressions of some inherent "human nature," but rather are conventions set up by society. "Man is the measure of all things" wrote the Sophist Protagoras, "of things that are, that they are; of things that are not, that they are not." This idea, too, had considerable impact when it was advanced, and even those who resisted it—a group including both ultra-conservative aristocrats and brilliant speculative philosophers—were evidently made uneasy by what we may call an intuitive plausibility that the suggestion had. Nineteenth-century scholars tended to assume that this new stress on "convention" as opposed to "nature" was the result of new "anthropology," namely, travellers' accounts, showing the Greeks how widely custom in fact varied from one culture to another. That may be part of the truth, but to me the most impressive evidence for the idea that society creates a world by arbitrarily setting up public standards lies no further away than the familiar market place. There, as anyone could see, an elaborate framework of convention—coinage, units of weight, standards of liquid measure, of grain measure, of nut measure, of roof-tile measure—was set up by fiat, and then used to give an operational definition of business ethics, fair trade, and honesty.

A third interaction is the continuing exchange of ideas between the natural sciences, the generalizations of philosophers, and the improved sense of mechanical logic of the craftsmen and gadget-makers. This began with Anaximander and continued through Roman times. We can now, as the nineteenth century could not, understand some passages written by Plato during his directorship of the Academy which seem to be illustrating cosmological myths and theories by reference to models. After an examination of the evidence, I have concluded that the reason Plato's passages seem to describe models *is that in fact this is what they are doing.* (The detail is presented in an article, "Plato and the History of Science," *Studium Generale* [1961], 520–527.) As was indicated in chapter four, above, since the inception of the "otherworldly" interpretation of Platonism by the Neo-Platonists, dating from the third century A.D., most philosophers and historians of science have assumed that Plato's Academy had no use whatever for experiment or observation. Instead, they were credited with an antiempirical program for pure theory construction. This assumption was shared by scholars who thought this a good thing and by those who thought it, as Eric T. Bell, the historian of mathematics, briefly states, "the worst disaster science has suffered." But if my reading of Platonic passages in the context we have been exploring is correct, Plato himself must have seen no incompatibility at all between looking for general cosmological equations and designing some type of model to provide analogues of sun, moon, and planetary relative motion. The supposed break between Platonism and any technological or empirical experiment and verification is a later interpretation of Plato, not a part of his own philosophy.

Finally, in the Roman world we seem to find what we have called "experimental theology," deriving from a combination of new ideas and new machine design. The reader will recall what the issues at stake in this experiment are. As we have said, the Stoic God was material—a "divine breath" (*pneuma*), penetrating and animating the universe. The Epicureans, carrying on the atomic theory as *their* model of scientific explanation, evidently thought they could show on experimental grounds that the God of Stoicism did not exist. For, the argument goes, a "fine spiritual breath" is *too* rare, too insubstantial, to move or animate anything made of more dense and solid matter. But the Stoics were able to offer a rejoinder, using steam or hot air as analogues of God's "spiritual power". A small ball could be shown supported by air alone as in Heron's "The Earth in Space"; a jet of steam could be made to support a sphere in the air; and a controlled steam jet could spin the *aeolipile* which Heron described.

I hope that my selection of items gives a fair sample of Western development of the combination of mechanical logic, a desire for novelty, and inventive ingenuity so brilliantly evolved that it has become a part of the taste and "common sense" of Americans in the twentieth century.

The Industrial Revolution marked the modern discovery of ways of harnessing new kinds of power to all the tasks that before had required slave labor or animals. New high rates of production in the West have created a new sort of economic democracy, in which a leisure and luxury beyond the reach of ancient aristocrats have become common property of almost all of our modern citizens, and potentially possible for everyone. Having discovered the ideas of applied science and application of new power sources, and got them clearly in mind, we have made good use of them. Steam and water power have been followed by gasoline and electricity, and these in turn promise to be partly supplanted by atomic power. Democracy has evolved along a line that is a compromise between the extreme of the Athenian lottery and the counterproposal by its critics to allocate every job to the citizen best qualified to hold it. Legislative procedure, too, has advanced. Representative government with a stable constitution has replaced the perpetual referendum of the Athenian Assembly; and we now have small but selected juries, instead of the huge but purely random ancient panels, with a gain in justice and impartiality. The change has been so great both qualitatively and quantitatively that we find it hard to re-create, in imagination, the life of the ancient Greek and Roman world from which we inherited the key ideas and techniques that made our progress possible.

However, in spite of this incredible advance in applied science and inventiveness, the interaction of ideas and machines poses modern counterparts of the questions—about life, the universe, honesty, and leisure diversion—that we have been exploring in their classical setting. We are still fascinated by the question of whether we can synthesize life chemically, or create machines that think electrically. And, if we can, opinion still will differ as to what the result shows about the nature of the human "self" or the "soul". We are still debating —on a much more complex and sophisticated level—what sort of cosmological model best fits the results of our observations of astronomical phenomena. Is the best model of our universe one that expands indefinitely, or one that pulsates, or one that remains in a steady state? And we wonder what relation holds between our familiar terrestrial geometry and the geometry of cosmic space? However great an advance the modern bank vault seems over the Mycenaean gate and lock, it is evidence that we are still working assiduously on the same problem of the automation of honesty. New power sources, as we have said, offer us a world with undreamt-of possibilities for labor saving, a world in which slavery would no longer make sense economically, even if it were not morally wrong. Toys for adults—from self-propelling automobile to the magic theater of television—are with us, too. Indeed, there are so many of these, and they are so attractive, that there is some concern for fear we may lapse into lives like those of the prisoners in Plato's Myth of the Cave, who spend all their waking hours watching the shadow play on the wall.

Human nature still retains its fondness for the surprising, its taste for ingenious mechanisms, that found an earlier expression in the nonindustrial but fascinating classical devices. Novelty pitchers, scout knives, souvenir spoons, carrot scrapers, cast-iron lawn ornaments, silver bar tools surround us with a world whose texture is shot through with threads of gadgetry. And so, as Joanna's hat design reminded me when I was writing the introduction to this book, in spite of many changes my conclusion is that human curiosity and ingenuity, from ancient Greece to modern America, are two qualities that have stayed about the same.

ANNOTATED BIBLIOGRAPHY

American School of Classical Studies in Athens. *The Athenian Agora, A Guide.* Rev. ed., Princeton, 1963. This, and the American School's Agora Excavations Picture Books, listed by author below, are essential sources for information about ancient Athenian everyday life, pans, weights, clocks, courts, and so on.

Apicius. *Artis magiricae.* With trans. by B. Flower and E. Rosenbaum. London, 1951. The only extant ancient cookbook. Roman, but with a dozen older Greek recipes. Gives an idea of the technical work in a classical kitchen.

Aristophanes. *The Clouds. The Birds.* Various English translations. The text of *The Clouds* edited by W. W. Merry (Oxford, 1894) has useful notes on the play.

Aristotle. *The Basic Works of Aristotle.* A selection from the Oxford trans., edited with an introduction by R. McKeon, New York, 1950. A compact one-volume collection; the editor's "Introduction" is particularly helpful to an understanding of Aristotle's over-all strategy.

──────. *The Constitution of Athens, and Related Texts.* Trans. with notes by K. von Fritz and E. Kapp. New York, 1950. An Aristotelian work recovered at the end of the nineteenth century. Our primary source for the amazingly complex procedures of election by lot and jury selection. (Also see S. Dow and M. Lang, below.)

──────. *The Works of Aristotle Translated into English.* Ed.

Sir David Ross. 13 vols., Oxford, 1908–1959. Cited above as "Oxford trans." Four volumes are particularly relevant to the present discussion.

De anima. In vol. III. This is Aristotle's analysis of the nature of "soul" and its relation to "body." A critical text with notes has been prepared by Ross (Oxford, 1956).

De caelo. In vol. II. Aristotle's general theory of the heavens and the "natural" motions of the terrestrial elements.

Metaphysics. Vol. VIII. This contains, in Book Lambda, Aristotle's brief description of his projected "nested ethereal sphere" model of celestial motion. For any detailed study, the text and notes in Ross's edition (2 vols., Oxford, 1924), are essential.

Nicomachean Ethics. Vol. XI. Book v is particularly interesting in its use of measure and ratio theory to define three types of "justice."

[Aristotle] or Pseudo-Aristotle. Various essays and notebooks by later members of Aristotle's school were included in and transmitted with his genuine works; they are included in the Oxford translation. Three of these are of interest for the present theme.

De motu. In vol. V of Oxford trans. A more mechanistic development of one side of Aristotle's psychology. Particularly interesting for the present study because a primitive "stimulus-response arc" is defended by a comparison of human behavior to that of marionettes.

Mechanica. In vol. VI of Oxford trans. The oldest book on mechanics; a notebook of mechanical problems, centering on the lever and the wheel.

Problemata. Vol. VII of Oxford trans. A large, miscellaneous "problem notebook." The only specific problems cited from this are the two questions about detecting loaded dice, but it is also helpful in understanding the kind of collection represented by the *Mechanica.*

Bowra, Sir M. *The Greek Experience.* London, 1957. Particularly good introduction to classical visual arts—architecture, sculpture, ceramics—with illustrations.

Britten's Old Clocks and Watches and their Makers. 7th ed., ed. G. H. Baillie, C. Clutton, and C. A. Ilbert. New York, 1956. A fascinating book for anyone who likes gadgets and ingenious machines. Has an excellent bibliography.

Brumbaugh, R. S. "Gadgets and Greek Philosophy," *Greek Heritage Magazine,* II (1965), 38–45. A brief outline of some types of relevance between archaeology and the study of classical philosophy.

———. "Plato and the History of Science," *Studium Generale,* IX (1961), 520–522. A discussion of possible use and design of astronomical or cosmological models in Plato's Academy.

———. *The Philosophers of Greece.* New York, 1964. An introduction to ancient Greek philosophy, with selected bibliography.

——— and Col. Paul H. Sherrick. "Pneuma and the Earth in Space," *Studium Generale,* XVII (1964), pp. 263–266. A reconstruction of Heron's "earth in space" machine, discussed in the text, and arguments for its identification as a Stoic cosmological model.

Bury, J. B. *A History of Greece.* New York, n.d. (Modern Library). An excellent one-volume history.

Chapuis, M. and A. Droz. *Automata.* Trans. A. Reid. Neuchatel, 1958. A definitive book on automata from Heron to the twentieth century, with a good opening chapter on ancient dolls and idols. Has an extensive bibliography and many excellent illustrations.

Claggett, M. *Greek Science in Antiquity.* London, 1957. A very good study of this aspect of ancient Greece. Probably better read after De Santillana or Price (see below).

Cumont, F. *Astrology and Religion among the Greeks and Romans.* Trans. J. B. Baker. London, 1912; reprinted New

York, 1960. Particularly interesting in connection with cosmology and cosmological models.

De Santillana, G. *The Origins of Scientific Thought.* Chicago, 1963. A lively study of the development of classical science, with interesting selected translations of source material.

Diels, H. *Antike Technik.* 3rd ed., Leipzig, 1924. An early study of Greek technology. Special topics treated include city planning, doors, and ciphers.

——, and W. Kranz. *Fragmente der Vorsokratiker.* 10th ed., 3 vols., Berlin, 1901. The standard collection of source material for early Greek philosophy. (See K. Freeman, below, for an English translation.)

Diogenes Laertius. *Lives and Opinions of Eminent Philosophers.* Trans. R. D. Hicks. 2 vols., Loeb Classical Library. New York, 1925. Gossipy, not very reliable scrapbook from the 2nd century A.D. Lives include Anacharsis, Thales, Anaximander, Pythagoras, Archytas, Plato, and Aristotle. The sections on the Stoics and Epicureans were taken from contemporary material and are reliable and important sources for these schools.

Dow, S. "An Athenian Lottery Machine," *Harvard Studies in Classical Philology,* 50 (1939), 1–34. An article that has become a classic; the final unravelling of the mechanics of Athenian jury selection and the mechanism of the *kleroterion.*

Forbes, R. J. *Man the Maker: A History of Technology and Engineering.* New York, 1950.

——. *Studies in Ancient Technology.* 8 vols., Leiden, 1955–63. Detailed investigations of ancient technological processes.

Vol. I: *Bitumen and Petroleum in Antiquity; The Origin of Alchemy; Water Supply.* 2nd rev. ed., Leiden, 1964.

Vol. II: *Irrigation and Drainage; Power (the story of water- and windmills); Land Transport and Road Build-*

ing; *The Coming of the Camel*. 2nd rev. ed., Leiden, 1965.

Vol. III: *Cosmetics and Perfumes in Antiquity; Food, Alcoholic Beverages, Vinegar; Fermented Beverages, 500 B.C.–A.D. 1500; Crushing; Salts, Preservation Processes, Mummification; Paints, Pigments, Inks and Varnishes*. 2nd rev. ed., Leiden, 1965.

Vol. IV: *Fibres and Fabrics of Antiquity; Washing, Bleaching, Fulling and Felting; Dyes and Dyeing; Spinning, Sewing, Basketry and Weaving*. 2nd rev. ed., Leiden, 1964.

Vol. V: *Leather in Antiquity; Sugar and Its Substitutes; Glass*. Leiden, 1957.

Vol. VI: *Heat and Heating; Refrigeration, the Art of Cooling and Producing Cold*. Leiden, 1958.

Vol. VII: *Ancient Geology; Ancient Mining and Quarrying; Ancient Mining Techniques*. Leiden, 1963.

Vol. VIII. [*Early Metallurgy, the Smith and His Tools, Gold, Silver, and Lead, Zinc and Brass*]. Leiden, 1964.

———, and E. Dijksterhuis. *A History of Science and Technology*. 2 vols., Baltimore, 1963.

Freeman, K. *Ancilla to Pre-Socratic Philosophy*. Oxford, 1948. An English translation of the direct quotations from the early Greek philosophers, from the collected edition of Diels and Kranz.

———. *Companion to Pre-Socratic Philosophy*. Oxford, 1948. Translation and paraphrase of the criticisms, paraphrases, and anecdotes about the early Greek philosophers, from the collection of Diels and Kranz.

———. *Greek City-States*. New York, 1950. Case studies of some of the city-states of ancient Greece, with some sympathetic appraisal of why the Greeks found this system great.

———. *The Murder of Herodes, and other Trials from the Athenian Law Courts*. New York, 1963. A selection of translations of extant samples of ancient legal rhetoric.

Gordon, C. H. "The Ancient Greeks and the Hebrews," *Scientific American* (February 1965), pp. 102–111. Particularly interesting in its treatment of the "Phaistos Disk" inscription.

Graves, R. *Greek Myths.* Harmondsworth, 1955.

Grace, V. R. *Amphoras and the Ancient Wine Trade.* American School Picture Books, #6. Princeton, 1961. Trade marks and jar designs figure in this reconstruction of one of the main commodities in classical commerce.

Greece. Hachette Blue Guides, Paris, 1957. Like the Hachette *Athens* Guide, a remarkably scholarly and comprehensive piece of work. Includes the sites on Crete, and a brief inventory of the exhibits in the Herakleion Archaeological Museum.

"Greece." *Life Magazine,* March 18–May 20, 1963. Excellent photographs with good text; G. De Santillana's article on science and philosophy (Part 3, April 22, 1963) is particularly interesting.

Greece. Edited by A. Eliot and the Editors of *Life Magazine.* Life World Library. New York, 1963.

Guthrie, W. K. C. *A History of Greek Philosophy: Part One.* Cambridge, Eng., 1963. A scholarly, judicious summing up of the earliest period of Greek philosophic thought. Very good bibliography.

Hamilton, E. *The Greek Way to Western Civilization.* New York, 1930. Very good general view of ancient Greek literature.

Hanson, N. R. "On Counting Aristotle's Spheres," *Scientia*, XII (1963), 223 ff. Includes a cross-section drawing of the complex mechanism of Aristotle's celestial mechanical model, and suggests that this includes too many "counteracting" components to actually function as Aristotle thought it would.

Harrison, E. B. *Ancient Portraits from the Athenian Agora.* American School Picture Books, #5. Princeton, 1960.

ANNOTATED BIBLIOGRAPHY 143

Head, B. V. *Historia Numorum.* Oxford, 1911. A standard reference work on Greek coins.

Heath, Sir Thomas. *History of Greek Mathematics.* 2 vols., Oxford, 1921. The standard work in English; in passing, Heath also treats the mechanical work of such figures as Archimedes, Pappus, and Heron of Alexandria.

Heron of Alexandria. *Opera,* I. *Pneumatica* and *Automata.* Ed. W. Schmidt. Leipzig, 1894.

———. *The Pneumatics of Heron of Alexandria.* Trans. for and ed. by Bennett Woodcroft. (Trans. by J. G. Greenwood.) London, 1851.

The Horizon Book of Ancient Greece. New York, 1964. An elegant folio of illustrations and well-chosen translations, constituting a very good sample of ancient Greek literary and graphic art.

Kahn, C. H. *Anaximander.* New York, 1962.

Kitto, H. O. *The Greeks.* Penguin Books. Harmondsworth, 1951. Particularly good, correct, and original in its presentation of the role of the Greek *polis* in the life of its citizens.

Lang, Mabel. *The Athenian Citizen.* American School Picture Books, #4. 2nd printing, Princeton, 1963. A fine collection of photographs, some of which are reproduced in my text, above, from the American School finds in the ancient Athenian Agora.

Liddell, H. G., R. Scott, and Sir H. Jones. *Greek Lexicon.* Oxford, 1940. The standard English lexicon; indispensable for locating references to technical equipment and processes in classical literature (e.g., *rhyton, gnomon, hyanon*).

McNeill, W. *The Rise of the West.* Chicago, 1964. An excellent book for locating ancient Greece in the broader context of world history. Some particularly interesting observations on agriculture, intercultural exchange, horsemanship, and warfare.

Neugebauer, O. *The Exact Sciences in Antiquity.* Copenhagen, 1951. A pioneering study, bringing out the importance of Babylonian computational astronomy for the history of Western science.

Pauly, A. F. and G. Wissowa (eds.). *Realencyclopedie* . . . 24 vols. in 49, with supplement of 9 vols. in 7. 1894–1959. The standard detailed reference work on classical antiquity.

Perlzweig, J. *Lamps from the Athenian Agora.* American School Picture Books, #9. Princeton, 1964.

Plato. *Dialogues. Letters.* Various translations and editions. A good recent English edition is *The Complete Works, including the Letters,* various translators, ed. by E. Hamilton and H. Cairns. New York, 1963.

A general bibliography of books and articles will be found in R. S. Brumbaugh, *Plato for the Modern Age* (New York, 1962), pp. 237–252. For Plato's most technical scientific work, the *Timaeus,* the Jowett translation which Hamilton and Cairns use should be supplemented by that of R. G. Bury (Plato, *Timaeus,* etc.), Loeb Classical Library, London and New York, 1929; or F. M. Cornford, *Plato's Cosmology* (the *Timaeus* translated with running commentary), London, 1937. Although A. E. Taylor's thesis that the *Timaeus* is only Plato's reconstruction of the science of an earlier period has not been accepted by scholars, his *A Commentary on Plato's Timaeus* (Oxford, 1928), is indispensable. Burnet's edition of the *Phaedo* (*Plato's Phaedo,* ed. J. Burnet, Oxford, 1911) has excellent notes; appendix II is my source of information concerning Glaucus of Samos. See "Yale College: Plato Seminar," below.

Plutarch. *Lives of Noble Greeks and Romans.* Dryden's trans., rev. H. Clough, Boston, 1857. The "Life of Marcellus" is the source of the account of Archimedes quoted in my text.

Price, Derek J. "An Ancient Computer," *Scientific American,*

200 (June 1959), cover, pp. 60–67. This gives illustrations and an account of the reconstruction of the "Antikythera machine."

———. "Automata and the Origins of Mechanism and Mechanistic Philosophy," *Technology and Culture,* V (1964), 9–23.

———. "Gods in Black Boxes," in Yale University, *Computers for the Humanities?,* New Haven, 1965, pp. 3–5.

———. *On the Origin of Clockwork, Perpetual Motion Devices, and the Compass.* Paper 6, Contributions from the Museum of History and Technology, U.S. National Museum Bulletin 218, Washington, 1959. Some of the results are described in Price's *Science Since Babylon.*

———. *Science Since Babylon.* New Haven, 1961. Some particularly valuable insights into the relation of astronomical models and the clock.

Robinson, H. S. "The Tower of the Winds and the Roman Market-Place," *American Journal of Archaeology,* XLVII (1943), 291–305. Photographs and fine description of the Roman Andronikos clock. (For more detailed drawings of the Tower of the Winds, see J. Stuart and N. Revett, *The Antiquities of Athens,* I, London, 1762, pp. 13–25, Plates I–IX.)

Sarton, G. *A History of Science.* Cambridge, Mass., 1962. An important work in the twentieth-century awakening of interest in the history of science. Tends to underrate and oversimplify Plato and Aristotle.

Schmidt, M. C. P. *Die Entstehung der Antiken Wasseruhr.* Leipzig, 1912. A detailed study, with illustrations, of early design of Greek water-clocks.

Seltman, C. *A Book of Greek Coins.* King Penguin Books. London, 1952. An attractive selection of illustrations of the coins which combined utility and fine art.

Seyffert, O. *A Dictionary of Classical Antiquities.* Trans. and rev. H. Nettleship and J. E. Sandys. London, 1876; reprinted New York, 1957. A compact treasury of information on various aspects of ancient life. (For example, this was my source for the game of *kottabos* and the vase illustration.)

Sparkes, B. A. and L. Talcott. *Pots and Pans of Classical Athens.* American School Picture Books, #1. Princeton, 1961.

Thompson, D. B. *Miniature Sculpture from the Athenian Agora.* American School Picture Books, #3. Princeton, 1959.

Thompson, D'Arcy W. "Games and Playthings," in *Science and the Classics* (Oxford, 1940), pp. 149–165. A nice summary of ancient toys and some games. One of these, involving putting a pin through a complicatedly twisted leather thong, suggests that such twisted thongs or strings could have been the motive power of classical automata.

Vitruvius. *The Ten Books on Architecture.* Trans. M. H. Morgan. Cambridge, Mass., 1914; reprinted New York, 1940. Book X, "On Machines" is a fine example of typically Roman "practicality," particularly when one compares it to Heron's works. One interesting device is the hodometer that is described in detail.

Yale College. Plato Seminar, 1964. "An Index of Plato's References to the Arts and Crafts." Mimeographed. New Haven, 1964. A proof that, however "illiberal" Plato sometimes said the arts and crafts are, he paid a good deal of attention to them.

Young, S. "An Athenian *Clepsydra*," *Hesperia* (1939), pp. 274–284. A six-minute official court room water-clock.

INDEX

Abaris, 27
aeolipile, 108, 134
air pressure, devices operated by, 10, 11, 20, 102
aither, 43
Ajax the Locrian, legend of, 124–125
alphabet, 16, 24, 25
"amazing things," 46, 50–51, 113
American School of Classical Studies (Athens), 39, 49, 71
Anacharsis, 22, 25
Anaximander of Miletus:
  concepts and inventions, 31–33
  cosmic models, 5, 31
  first map, 32, *ill.* 33
  sundial (gnomon), 70
anchor, Anacharsis' invention of, 25
Andronikos of Rhodes, 95
Apology (Plato), 56
applied science, ideas of, 2, 3
Archimedes of Syracuse, 3, 75–83, 84
  death of, 82, *ill.* 91
Archytas, 29, 77
Aristotle, 5, 12–13, 19, 40, 41, 47, 51, 56–58
  "celestial mechanics," 41–45
  Constitution of Athens, 63–64, 71
  cosmic model, 43–45

Aristotle *(cont.)*
  Poetics, 73
  psyche thesis, 27, 56–57
astronomy:
  Aristotle's "celestial mechanics" concept, 41–45
  computational, 73, 97
  mechanical explanation of atomic theory, 36–37
  models, 3–4, 33–34, 82–83, 95
Athenian Agora:
  map, *ill.* 39, 85
  relics from, 49, 71, 86, 92
  town clock, 6, 68, 72, *ill.* 90, 95
Athens (*see also* automation, of honesty), 7, 8, 10, 46
  lottery in, 59–67
  map, 39, 85
  mechanical clock time in, 6, 70
atomic theory, 5, 36, 55, 58, 131–132, 134
Automata (Chapuis and Droz), 17
Automata (Heron), 84, 113, 114–117
automation, 3, 5, 26–27, *ill.* 76, 100
  Heron's designs, 44, 57, 93–104, 113–129
  of honesty, 7–8, 15, 18, 19, 61–63

INDEX

automation (*cont.*)
  religious, 7, 8–10, 19–20, 74, 97, 100–102
  toys and theaters, 50–55, 113–129
axle-winding patterns, *ill.* 57, 114–120

Babylonia, 42
ballot box, Athenian, 132
Bell, Eric T., 134
Bohr, Niels, 34
brace and bit, 23
Broneer, O., 30
Burnet, John, 26
burning glass (mirror), *see* lenses
business, regulation of, *see* commerce, standardization of

causality, Aristotle's concept of, 57
"celestial mechanics," of Aristotle, 41–45
change, Thales' concept of, 5, 32
Chapuis, M., 17
Chrysippus the Stoic, 105, 107
clocks, 68–73
  klepsydra (small water), 6, 71
  water (town), 6, 68, *ill.* 69, 72, *ill.* 90, 95
Clough, H., 77
coins, 126
commerce, standardization of, 6, 7, 19, 60, 68, 132–133
commode, child's, 49, *ill.* 92
compass, magnetic, 27
computer, astronomical, 73, *ill.* 96, 97–98
conservation of design, law of, 51
Constitution of Athens (Aristotle), 63–64, 71
cooking devices, 47–49
cosmology:
  Aristotle's models, 41–45
  models, 5–6, 31, 34, 36

courtroom time, 72
creation, Plato's model of the world-soul, 38
Crete, Minoan culture in, 15–21
Crito (Plato), 56
Ctesibius, 52–53, 54, 84, 95, 98

Da Dondi clock, 70
Daedalus, 22, 23–27
Delphic knife, 12–13
Democritus, *psyche* thesis of, 27
De motu, 57, 138
dice, 24, 73–74
Discovery of the Mind, The (Snell), 55
doors, automatic, 10
  alarm bell, 10, 41
  Heron's designs, *ill.* 44, 101–102
Droz, A., 17
Dryden, John, 77

earth in space theory, 106–107, 134
  Heron's model, *ill.* 87
eclipse predicted by Thales, 31
engines, steam, *see* steam power
Epicurean school, 45
  pneuma theory, 105–106, 107, 134
eraser, stylus, 29
Euclid, 74
Eudoxus of Cnidos, 77
  "save the phenomena" solution, 40–41, 43
Euthyphro (Plato), 56

faucet, 9, 100, 101
formalism:
  Pythagorean concept of, 35
  versus mechanism, 37–38, 41–45
fountain, Heron's design, *ill.* 17, 111

INDEX

gameboard, 16, 19
games, see kottabos
gearing:
  computer, ill. 96
  Heron's design, 98, 119
  right-angle, 9–10, 11, ill. 51, 103, 105
Gemini II, 14
geometry, Archimedes' designs based on, 75
Glaucus of Samos, 35
grills, 47, ill. 86

Hanson, Norwood R., 45
heliocentric models, 38
Hephaistos, 39, ill. 92
Herakleion Archaeological Museum, 15
Heron of Alexandria, 9, 52–53, 54, 84–112
  Automata, 84, 113, 114–117
  automated theater designs, see theater
  automatic doors design, ill. 44, 101–102
  circular motion of an automation design, ill. 57, 117
  control wheel, 54–55
  "Dutch-treat" wine vessel design, 36, 111–112
  Earth in Space model, ill. 87, 106–107, 134
  fountain design, ill. 17, 111
  gadgeteer's toy design, ill. 72, 112
  lamp design, ill. 62
  lightning-effect machine, ill. 128
  pneuma theory, influence on, 106–108
  Pneumatica, 2, 84, 97, 98, 106, 113, 119
  revolving platform design, ill. 48
  sound-effect machine, ill. 28

Heron of Alexandria (*cont.*)
  steam engine, 4, 108
  steam power, use of, 4, 10, 108–109
  theory of vacuum applied to siphon, ill. 24, 97, 98, 102
Hiero of Syracuse, 76, 77–78
Hippodamas, 35
hippopede, see "horse-fetter"
honesty, automation of, see automation
"horse-fetter" (*hippopede*), 40, 120
human behavior, see puppets

Icaria Island, 6, 24
Icarus, 23–24, 25

jury selection, 8
  by lottery, 59, 61, 65–67

Kamares, 18
King Must Die, The (Renault), 23
kleroterion, see lottery machine
knives, 12–13
Knossos, 16, 18, 19
kottabos, game, ill. 13, 49–50

labor-saving invention, 2–3, 7, 8–9, 74, 97, 101
labyrinth, origin in palace of Minos, 24
lamp:
  Heron's design, ill. 62, 110
  relief from, ill. 92
lathe, 23
lenses, burning glass and magnifying, 16, 18–19, 82, 83
Leucippus, *psyche* thesis of, 27
lever, Archimedes' theory of, 77
Levinson, Ronald, 44
libation jars, 16, 19–20, 70, 101
Life of Marcellus (Plutarch), quote, 77–82

150   INDEX

lightning effect, Heron's device, *ill.* 127–128
lock and key, 29
lottery system, 59–67
   kleroterion, machine, *ill.* 66
   political relevance to Athenian democracy, 4
   tickets, *ill.* 89

magic, (*see also* miracles), 21, 23, 25, 26
   "dart" of Abaris, 27
magnifying lens, *see* lenses
many-purpose gadget, 12–14
maps:
   Anaximander's first, 32, 39
   Athens, Agora and environs, 33, 85
Marcellus, 75, 77–82
Marcianus manuscript, 104
mathematics, 74
   Aristotle's formalism versus mechanism concept, 41–45
   as basis of music, 35
   Pythagorean models, 34–35
matter, Thales' theory of, 32
measures (*see also* weights and measures), *ill.* 88
Mechanica, 9, 73, 74, 84, 100, 138
Meton, sundial of, 71
Milesians, 31, 34, 42
Minoan culture, 15–21
Minos, 23–24
miracles, temple devices performing, 9–10, *ill.* 11, 41, 101–102, 103–105
models:
   Anaximander's concept of, 31, 32–33
   Aristotle's cosmic, 41–45
   astronomical, 3–4, 33–34, 82–83, 95
   cosmic, 5–6, 31, 34, 36
   Heron's Earth in Space, *ill.* 87

models (*cont.*)
   in Plato's Academy, 37–38
   role in Greek astronomy, 3, 38
   Sherrick-Brumbaugh (universe), *ill.* 87
   science and model makers, 31–45
motion:
   astronomical models, 33–34
   cosmic, 42–45
   planetary, 35, 39
   psyche as the cause of, 5, 27
"movable-type," 15–16
music, mathematics as basis of (Pythagoras), 35–36
Mycenaean culture, 21, 26
Mykonos, 125
"Myth of the Cave" (Plato), 51–52

natural science, 133
   Thales' concept of, 5, 32
Nauplius (play), Heron's designs for, 54, *ill.* 99, *ill.* 103, 123, 124–127
Nestor's cup, 21
neura, 53

On Architecture (Vitruvius), 94

Palamedes, 22, 24–25
panpsychism, 58
Parmenides, poem, 29
pencil, stylus, 29
Phaedo (Plato), 56
Philo of Byzantium, 84, 122, 127
Philosophers of Greece, The (Brumbaugh), 31, 35
pitchers:
   "Mother Goddess," 16, 18
   novelty, 49, 98, *ill.* 86
Plato, 12, 35, 108, 133–134
   Academy, 37, 40
   analogy of puppets and man, 56

Plato (cont.)
  marriage lottery, 61
  model of the world-soul, 37–38
  "Myth of the Cave," 51–52, 53
  Socratic dialogues, 56
  Theaetetus, on clock watching, 72, 73
  Timaeus, 38, 40
plumbing system, 16
Plutarch, 84
  Life of Marcellus, quote, 77–82
pneuma theory, 105–108, 134
Pneumatica (Heron), 2, 84, 97, 98, 113, 119
Pnyx, 71, 85
Poetics (Aristotle), 73
Pots and Pans of Classical Athens (Sparkes and Talcott), 49
potter's wheel, 25
power sources, uses of, 2, 4, 83, 97, 114
Price, Derek, 71, 95, 97, 108, 113
Problemata, 73, 138
Protagoras, 132
psyche, see soul
Ptolemy, 73
puppets and marionettes, 50–58
  analogy of human behavior and, 46, 56, 58
  Heron's designs, *see* theater
Pythagoras, 27, 31, 34–35

quartz lenses, 16, 18–19

rattles, baby, 29–30
religious automation, 7, 8–10, 19–20, 74, 97, 100–102, 103
Renault, Mary, 23
Republic (Plato), 61
revolving platform, Heron design, *ill.* 48

sacrificial water dispenser, *ill.* 7, 9, 97, 100–102

"Sailing to Byzantium" (Yeats), 17
Sambursky, S., 105
"save the phenomena" solution, of Eudoxus, 40
Schmidt, M. C. P., 104, 111
science and model makers, 31–45
Scientific American, 97
Scottish dirk, 13
Scythians, 22, 25
seasons, measure of, 70
secret ballot, 62–63, 67
self, Socrates' search for, 55–56
self-motion, Plato on soul and, 56
self-moving automata, see automation
shadows, effected in puppet theater, 52, 53
Sherrick, Paul H., 10, 87, 106
siphon, Heron's theory of, *ill.* 24, 97, 98, 102
slot machines, *ill.* 7, 9, 97, 100–102
Snell, B., 55
Socrates:
  analogy of puppets and man, 55–56
  execution of, 56
  psyche thesis, 55, 56
Sophists, 132
soul, *psyche*
  Aristotle's concept of, 56–57
  atomic theory and, 56
  Plato's concept of, 38–39, 56
  Socrates' concept of, 55, 56
  as source of motion, 5
  in Thales, 5, 32
sound effects, Heron's designs, 11, 41, 96, 101, 103–104, 111, 123
Sparkes, Brian A., 49
starting gate, 30
statues:
  of Daedalus, 23, 25–27
  souvenir, 49

steam power:
  engine, 4, 10
  Heron's use of, 4, 10, 48, 108–109
Stesilaus of Athens, 12, 29
Stoic Physics (Sambursky), 105
Stoics, 45
  pneuma theory, 105–106, 107, 108, 134
Studium Generale, 107, 133
sundials, 70–71

Talcott, Lucy, 49
taxation, 18
Teubner texts, 110, 111, 123
Thales of Miletus, 5, 31–32, 70
Theaetetus (Plato), 72, 73
theater, automated, 27, 50–55, 114, 120–129
  Heron's designs (first type), *ill.* 80, 93, 120–121
  Heron's designs (2nd type, *Nauplius*), *ill.* 99, 103, 123–127
  sound-effect machine, *ill.* 28, 122
Theseus and Ariadne legend, 23
tiles, standard, *ill.* 89
Timaeus (Plato), 38, 40
time (*see also* clocks):
  mechanical time, 6, 69–71
tools, 19
"Tower of the Winds," clock, 6, 68, 72, *ill.* 90, 95
town-clock, *see* clocks
toys, 19, 30, 47, *ill.* 92, 98
  Heron's designs, *ill.* 72, *ill.* 108, 112

toys (*cont.*)
  windup, 50, *ill.* 51, 113–120
  "toys for adults," 4, 10, 16, 110
tripod, of Glaucus, 35–36
triskeli, 126
trumpets, automatic, 10, *ill.* 20, 101, 111

vacuum, Heron's use of, *ill.* 24, 97, 98
vase paintings, 49–50
Vitruvius, 95
  On Architecture, 94
"vortex" model, 36, 45

war machinery, 75, 77–82, 91, 95
water-clock, *see* clocks
water-drinking toys (Heron), *ill.* 24, 98
water pump, 82, 83
weapons, 12
weather, "Tower of the Winds," 6, *ill.* 90, 95
weights and measures, standardization of, 8, 19, 60, 68, *ill.* 88
wheel and axle, 114–120, 127–129
  Heron's designs, *ill.* 57, 118
wheels, 9–10, 11, *ill.* 54, 74, 103
wine vessels:
  "Dutch-treat," *ill.* 36, 111–112
  psykter, 46–47, *ill.* 86
wings of Daedalus, 24, 25
Woodcroft, Bennett, 108, 110–111
writing, 15, 16–18, 29

Yeats, W. B., 17